CHERNOBYL AND NUCLEAR POWER IN THE USSR

CHERNOBYL AND NUCLEAR POWER IN THE USSR

David R. Marples
Research Associate, Canadian Institute of Ukrainian Studies
University of Alberta

St. Martin's Press   New York

First published in the United States of America in 1986

Printed in North America

ISBN 0-312-00414-1

Library of Congress Cataloging-in-Publication Data

Marples, David R.
  Chernobyl and nuclear power in the USSR.

  Bibliography: p. 197
  Includes index.
    1. Nuclear industry—Soviet Union.   2. Nuclear
power plants—Ukraine—Chernobyl—Accidents.
I. Title.
TK9085.M37   1986   333.79′24′0947   86-42967
ISBN 0-312-00414-1
ISBN 0-312-00457-5 (pbk.)

Contents

Acknowledgements vii
Introduction ix
Chapter One: A Chernobyl Diary, 28 April–14 May 1986 1
Chapter Two: Soviet Energy in the 1980s 37
Chapter Three: Nuclear Energy Development in Eastern Europe 51
Chapter Four: Ukraine in the Soviet Nuclear Energy Programme 71
Chapter Five: Safety in the Soviet Nuclear Power Industry 95
Chapter Six: The Chernobyl Disaster 115
Chapter Seven: After Chernobyl 153
Epilogue..181
Notes 185
Selected References 197
Appendices 203
Index of Personnel 209

Acknowledgements

I am indebted to a number of individuals and institutions who have assisted me over the past few months. The book was completed with the financial assistance of the Canadian Institute of Ukrainian Studies (CIUS), University of Alberta. I would like to thank Bohdan Krawchenko of CIUS for his help and encouragement, and Khrystia Kohut for her assistance in typing various materials. I am also indebted to the staff of Radio Liberty, Munich, for access to research materials during 1984-5. The writer also expresses gratitude to the staff of the Whiteshell Nuclear Research Establishment (Atomic Energy of Canada Limited), Pinawa, Manitoba for assistance with some of the technical data and information about the nature of the nuclear disaster: Bob Dixon, David Torgerson, Heiki Tamm, A. Soonawala, J. Hillier, Joseph Borsa, D.F. Dixon, Marsha Sheppard, L.H. Johnson, Gary R. Simmons and R.S.W. Shewfelt. The opinions expressed, and any errors therein, are mine.

A number of individuals also helped in various ways with advice and materials, including Jerry Magennis, Michael Klefter, Nadia Diuk, Jerry Holowacz, Peter Rolland, William Zuzak, Myroslav Yurkevich, Jurij Dobczansky, Vladimir Socor, R. Busch, Paul Goble, George Stein (*Los Angeles Times*) and Tassie Notar (*Canadian Broadcasting Corporation*). The interest and support of the Ukrainian community of North America has also been warmly appreciated.

My principal debt, however, is to my friend and former colleague at Radio Liberty, Roman Solchanyk, without whose unflagging assistance with research materials, advice and information, this work would not have been completed.

Finally, my wife Lan helped me throughout my labours, in preparing the index, proofreading the text, and by suffering my long hours and frequent absences with great patience and understanding.

David R. Marples

Edmonton, Canada,
August 1986

Introduction

The Chernobyl nuclear disaster marks a watershed in the history of the world nuclear power industry. The accident in the northern reaches of Ukraine received world attention as soon as the radiation cloud that resulted drifted over Scandinavia. The Soviet authorities, who had not divulged any news about the accident two days after the event slowly began to release information. In the meantime, Western press agencies began to speculate, sometimes wildly, about what had taken place and the number of casualties that may result. At the time of writing, there have been no firm conclusions about the causes of the accident, although it seems clear that it was a result of both technological problems and human error.

This book analyzes the Soviet nuclear power industry. In origin, it predates the Chernobyl accident, but inevitably its format has been determined by that event. It seeks the answers to several pertinent questions. First, why have the Soviet authorities committed themselves so heavily to the development of nuclear energy, given that the USSR is one of the only two countries in the world that can lay claim to a energy self-sufficiency? Second, has the speed with which the industry is being developed led to the neglect of the safety of citizens and the environment? Is there, for example, a well-documented history of neglect and general safety problems at Soviet nuclear power plants?

Third, is Chernobyl representative of the nuclear power industry in the USSR? Are there nuclear plants in a similar condition, facing similar dilemmas with supply and a lack of qualified and a surplus of dissatisfied workers? If so, does the Soviet industry constitute a living danger for the world at large? Finally, what will be the long-term effects of the accident, both on the immediate environment, for agriculture, and for the Soviet energy programme? Can the build-up of nuclear power continue under the new and difficult circumstances both in the Soviet Union and in Eastern Eugope, where a Soviet-run plan has been implemented?

In beginning this study, the author decided to limit himself largely to Soviet source materials. In his view, it is possible to glean enough information therein to make adequate conclusions about various facets of the industry. It should be stated at the outset that the object is not to judge or condemn the USSR, or to make any kind of political comment, but to analyze an industry that has remained shrouded in secrecy since its inception in the USSR in 1954. Moreover, the author's sphere of expertise pertains to the Soviet economy rather than nuclear physics. Thus in some sections, where specialized technical information is required, he has been obliged to rely on the information of scientists in the field.

The volume cannot provide a definitive account of either the Soviet nuclear power industry or of the Chernobyl disaster in particular. Both are still in progress. Its aim is rather to elucidate some of the important issues; to show how the disaster affected Soviet thinking; and to look at its impact on the lives of ordinary citizens.

The author is also of the view that one cannot discuss Chernobyl without some understanding of its setting: the Ukrainian SSR and the role assigned for Ukraine in the Soviet nuclear programme, both for domestic and East European supplies of electricity. Ukraine's nuclear power build-up has taken an extreme form. Immense pressure has been placed on local officials to comply with ambitious development plans. Ukraine has remained one of the most important economic regions of the USSR , but in several key spheres, its industries have stagnated or declined in output: coal, steel and chemicals being the most important. For Ukrainian officials, nuclear power represented a way out of an impasse, a passport to an enterprising future.

Consequently, while this book comprises a study of the Soviet nuclear power industry in its entirety, the emphasis is on the Ukrainian scene. Ukrainians in the West have catalogued Chernobyl as another chapter in a sad twentieth-century history that includes a man-made famine in 1932–3, the Stalinist purges of the 1930s, and many of the major conflicts of the German-Soviet war of 1941–5. At no point therefore should it be forgotten that the disaster took place on Ukrainian territory, and moreover, within the vicinity of an old historical town dating back to the late twelfth century: Chornobyl. Because of the publicity accorded to the Russian form of the name, this work uses "Chernobyl" throughout. The same also applies to Kiev and to the Dnieper River. In all other instances, however, the names of Ukrainian officials, town and villages are given in their Ukrainian form.

The book is divided into seven chapters. The first looks at one of the signifcant episodes of the whole affair, namely how the Soviet authorities released information about the accident over the first three weeks. The author's view is that while an apparent reluctance to release hard information can hardly be termed untypical for the Soviets, it shows nonetheless how much was at stake for the authorities in the economic sphere. Whether reports took the form of a news report from the site itself, a speech by Mikhail Gorbachev, or an *Izvestiia* correspondent's denunciation of the West followed by a listing of "accidents" at nuclear power plants outside Eastern Europe, the goal was the same: to avoid prejudicing the future of nuclear power in the USSR.

Chapter Two looks at the energy questions facing the Soviets, and why planners feel that the country can no longer rely on supplies of coal and oil to fuel the Soviet power industry. Chapter Three focuses attention on Eastern Europe and its part in the overall nuclear energy plan, especially its links with the Ukrainian SSR in the Council for Mutual Economic Assistance (CMEA). It examines the immediate impact of the Chernobyl disaster on nuclear power in Eastern Europe.

Chapter Four examines nuclear power development in the USSR in the Tenth and Eleventh Five-Year Plans (1976–80 and 1981–85 respectively) and looks at the prospects for the Twelfth Five-Year Plan and to the year 2000. It outlines Ukraine's part in this plan and provides a history and picture of recent developments at the individual stations in Ukraine. The object is to show that there were many common dilemmas at the time of the accident, including supply problems, defective materials, severe labour problems, alcoholism, and the shortage of qualified personnel. Chapter Five applies this scenario to the USSR as a whole, in somewhat broader perspective by analyzing the question: are Soviet nuclear power plants inherently unsafe?

The final two chapters focus on the accident itself and its aftermath. They portray the background of the Chernobyl station and the details of the first hours of the accident. An analysis is provided of the evacuation procedure, the clean-up campaign and the political repercussions of the tragedy. The extent and possible effects of radiation are also encompassed, while background information is provided on some of the officials leading the campaign "to eliminate the consequences of the accident" (to use Soviet parlance). Lastly, the author looks at the future of the nuclear industry and the impact of Chernobyl from a world dimension. Nuclear power cannot be confined within state borders and it if is to continue as a leading energy source, then it seems clear that even a totalitarian society must assent to some form of international control. The question is how thorough and how complete can that scrutiny be?

In 1984-5, the author was an employee of the U.S. radio station, Radio Liberty, in Munich. While there he was able to collect a copious amount of information on the Soviet nuclear power industry, culled from journals, newspapers and staff resources. As a repository of Soviet sources and of current information, Radio Liberty and its sister station, Radio Free Europe, are probably unmatched in the Western world. He relied also on the extensive newspaper resources of the Canadian Institute of Ukrainian Studies at the University of Alberta. Technical assistance was provided by employees of Atomic Energy of Canada Limited.

A Chernobyl Diary, 28 April–14 May 1986

28 April, 1986

On 28 April, at 2100 hours, *Radio Moscow* made a terse announcement:

> An accident has occurred at the Chernobyl nuclear power plant—one of the atomic reactors has been damaged. Measures are being undertaken to liquidate the consequences of the accident. Those affected are being given aid, and a government commission has been created.

The announcement came about eight hours after Swedish officials discovered high levels of radiation on the monitoring equipment at a nuclear power plant near Stockholm. It set off speculations in the West that one of the reactors at the Chernobyl plant had suffered a partial or total meltdown, that casualties must have been heavy—around 2,000 according to one UPI report—and that an important area of Soviet farmland had been contaminated for years to come.

Western speculations about the extent of the damage caused by the Chernobyl accident—while difficult to justify—were to some extent unsurprising. For the first time, the Soviet authorities had conceded that an accident had occurred at one of their nuclear installations. It was not, however, the first incident at Soviet nuclear plants to be reported in the West. Over the previous three decades, there had been a variety of accounts of incidents at Soviet nuclear plants, beginning with a major disaster: an apparent explosion of a nuclear waste dump near Cheliabinsk and Smolensk in 1958–9, which wiped out several villages and contaminated enough lakes and vegetation to make the entire area uninhabitable.

In the case of Cheliabinsk and more minor incidents that followed, the Soviet response to the Western reports had always been one of silence. On the other hand, there have been innumerable statements about the utmost reliability and safety of currently existing Soviet nuclear energy installations. On a *Radio Moscow* broadcast of 1 April 1985, for example, the science and engineering programme hosted by Boris Belitsky replied to a series of questions posed by a British listener from Bournemouth. One of the listener's questions concerned the alleged links of nuclear power with cancerous diseases and asked what was the USSR's position on this "and the associated safety aspects?" Belitsky replied that:

> Soviet experts are of the strong opinion that the development of nuclear power is essential to meet the country's growing electricity needs. As for the safety aspects, there's an equally strong opinion that properly designed and constructed nuclear power stations are quite safe. In fact, nuclear power stations in the Soviet Union have a very good safety record. Let me point out that there has not been a single major accident at a Soviet nuclear station detrimental to human health, let alone involving any fatality.

Belitsky's answer was not unequivocal because he used the adjective "major" implying that there may have been minor incidents in the past. But otherwise his response was fairly typical of the Soviet authorities as a whole.

In February 1986, *Radio Kiev* was extolling the safety mechanisms of the Chernobyl plant in apparent ignorance of the criticisms made of the plant elsewhere in the Soviet press. The broadcaster maintained that the plant was foolproof, and that the air around was totally uncontaminated. As for the plant reservoir, it was being used as a major fish-breeding source. After a quarter of a century of exploitation, the broadcaster continued, there had not been a single incident at a Soviet nuclear plant. This may have been bravado, but may also have reflected Soviet complacency about nuclear safety.

In view of the above comments, it is clear that the initial effect of the Chernobyl disaster was psychological. Such an accident, in theory, could not take place. At the same time, the Soviet leaders were perplexed and uncertain about what stance to take both before their own citizens and before the world at large. Their first reaction appears to have been to do nothing; when a nuclear steam generator had broken down at the Rovno plant, 150 miles west of Chernobyl in 1981, such a silence, from the Soviet perspective, may have been justified. The generator was repaired and the Soviets never officially admitted a problem. The 28 April statement reflects in its brevity this same attitude. A Swedish protest had forced a statement, but it was a statement that revealed little.

In fact, there have been precedents to the 28 April statement that illustrate that it was very much a standard comment. Here is one example from the Soviet coal-mining industry, in which industrial accidents are far from uncommon:

> On 10 August, a methane gas explosion occurred at the Molodhvardeiskaia coal mine in the Ukraine, resulting in casualties. The Soviet and the Ukrainian governments have undertaken measures to aid the injured and the families of the dead miners, and to eliminate the consequences of the accident. (*TASS*, 11 August 1979.)

The difference between a mining accident in the Donbass and the Chernobyl disaster, however, is that the latter increased in its scale and in its enormity with every passing day. Moreover, whereas the Soviets could internalize a mining accident, Chernobyl became a worldwide affair immediately as a result of the radiation cloud that moved northwest across the Scandinavian countries at the outset.

Faced with an international crisis that only became accentuated as a result of a typical Soviet silence in the first days of the accident and Sweden's angry protests, the Gorbachev regime adopted the policy of releasing information to the public in stages. One can surmise that at some point immediately after the Soviet leadership received news of the accident, it made two decisions: first, to hold back some of the details about the accident, in order to lessen the impact of an extraordinary event; second, to mount an intensive propaganda campaign illustrating analagous and allegedly more frequent nuclear accidents that had occurred in the West, evidently to place Chernobyl "in perspective" and to prevent any sort of panic among Soviet citizens, who would be aware of some details sooner or later from foreign radio broadcasts.

29 April 1986

The day after the first official announcement of the Chernobyl accident, the Moscow press continued its silence about the event. The USSR Council of Ministers statement was, however, reported in the major Kiev-based newspapers, but not in positions of prominence. *Pravda Ukrainy* placed it at the foot of page three, beneath an article about two sickly pensioners who were trying to acquire a telephone in their homes. *Robitnycha hazeta* gave the statement a similar location, this time below the Soviet soccer league tables and reports about a chess competition.

In the evening *TASS* made a second announcement which gave the following details: an accident had occurred at the Chernobyl nuclear plant, which is located 130 kilometres north of Kiev; a Government Commission, headed by Borys Shcherbyna, Deputy Chairman of the USSR Council of Ministers, had been established, which included "heads of ministries and departments." The accident had occurred "in one of the areas of the fourth power-generating unit and resulted in the destruction of part of the structural elements of the building housing the reactor." *Two people had been killed during the accident.* The remaining three reactors had been shut down, and the residents at the reactor site [Prypiat] and three neighbouring population points had been evacuated.

While only a little more revealing than the first official Soviet statement, it was becoming possible to ascertain which villages had been evacuated. Along with the nuclear plant city, Prypiat, which had a population of about 25,000, two of the settlements in question could have been Kopachi (population 1,024), 12 kilometres north of the raion capital of Chernobyl, and the village of Paryshiv (population 1,046), which is a mere seven kilometres from Chernobyl. The third village was probably Cherevach (population 630), the location of a collective farm that specializes in dairy cattle.

The Soviet authorities would have been aware that given the paucity of hard information in the first two statements that a hungry Western media would seize on the figure of two dead, the only information of substance that had been thrown in their direction. U.S. satellites were already taking pictures of the damaged reactor that according to some writers indicated a higher toll. The U.S. arms negotiator Kenneth Adelman, for example, dismissed the figure of two dead as "preposterous," thereby fuelling a propaganda war over the number of casualties that was to continue for some time.

As the official information filtered out, the staff at *Radio Moscow*'s World Service were issuing a series of conflicting broadcasts. At first, it was decided to utilize the Chernobyl affair to make a political point about the need to eliminate nuclear weapons, which reflected Soviet General Secretary Gorbachev's current political stance:

> Drastic measures are being carried out to guarantee the power reactor's reliability and safety. Nevertheless, our observer notes, as this accident and many others at nuclear power facilities in Western countries show, the application of nuclear power for peaceful purposes can be dangerous—and it is all the more obvious what a horrible threat nuclear weapons and their testing pose to all nations.

While Western listeners, some of whom had relatives in Ukraine, were anxiously awaiting further information about the extent and nature of the accident, *Radio Moscow* was indulging in petty politics. Were the Soviet authorities still unaware of the full scale of the accident?

The reporting continued its unusual course later the same day, turning on the traditional adversary, the United States. Alluding to the "major accident" at the Three Mile Island nuclear power plant in Pennsylvania in 1979, commentator Iurii Zolton declared that:

> that accident was caused by the criminal neglect of the plant's owners for the basic safety measures, which resulted in a discharge of radioactive substances into the atmosphere and a great deal of damage to the health of the local residents. Many of them are still suffering from exposure related diseases.

Zolton went on to attack the alleged buildup of nuclear weapons in Western countries. Implicit in these comments is the attitude within the Soviet leadership that Chernobyl was first and foremost a political setback that had to be rectified with a propaganda message.

At 2110 hours Moscow time, however, the radio station did an almost complete volte-face. A commentary by Igor Pavlov was a masterpiece in moderation, which at the same time carefully defended the record of the nuclear power industry. Given that it followed Zolton's angry statement so closely, it could only have been intended to lessen the import of the previous message:

> Since atomic power stations were introduced in the late 1950s there have been a series of major or minor accidents in France, Great Britain, the United States *and other countries*. According to the experts, the most frequent cause of radiation leaks is the overheating of the reactor's core. Incidentally, in all such accidents fewer people have been killed or injured than in several aircraft crashes over the same period and nobody has ever suggested that as a result all flights be terminated and people stop flying.

Cleverly, Pavlov informed listeners of a possible cause of the Chernobyl accident, and he went on to make comparisons with road accidents and the Space Shuttle *Challenger* catastrophe earlier in 1986, which had not stopped the U.S. space programme. "Major breakdowns at nuclear power plants," he concluded, referring specifically to Three Mile Island and Chernobyl, "call for a greater degree of international co-operation rather than an immediate shut down of all such facilities."

Pavlov appeared to be taking the more sensible line that a nuclear accident is an international rather than a domestic event. Yet even this broadcast hardly addressed the reality that was taking place. While Pavlov was speaking, Polish television was broadcasting a communique issued by

5

the Polish Government Commission for the Assessment of Nuclear Radiation and Preventative Measures, which noted the increase in the intensity of active iodine in the air, the sort of increase that "could be harmful to health if it were to occur over a lengthy period."

As a result, while the Soviet news services were indulging in polemics, the neighbouring Polish government was recommending that the populace should not consume milk from cows fed on green fodder, and that only milk from cows fed on dry fodder would be for sale. Further, the Polish Health Service was administering a single iodine preparation for babies and children in the northeastern regions of the country (which were most affected by radiation). The television programme also pointed out that "the Minister of Health and Social Welfare draws attention to the absolute necessity of washing all spring vegetables before eating." These measures were simple enough, but they appeared at this time to have been neglected by the Soviet government in the first days after the accident. The most charitable explanation is that the leaders were in a state of disarray, unsure about the extent of the contamination of their own citizens, and debating about what sort of face they should show the world and Soviet people.

In Hungary, which was not yet affected to the same degree as Poland by the radiation cloud, Budapest television was discussing the nature of the Chernobyl plant. Declaring that "we have tried all day to get more information," the reporter announced that the plant had four reactors, each with an output of 1,000 megawatts capacity "of the so-called single circuit type…an older model which is no longer manufactured."

The East European countries, which traditionally reflect the Moscow line in their broadcasts, were trying at least to provide their citizens with hard information. On this same day in the city of Kiev, which has a population of over 2.4 million, local officials and foreign students were declaring that "life was normal." (*Reuter.*) Foreign embassies were trying in vain to get more information from Soviet officials.

30 April 1986

On 30 April, both *Pravda* and *Izvestiia* allotted small inside columns of the newspapers to Chernobyl, in which they repeated the brief text released by *Radio Moscow* two days earlier, from the USSR Council of Ministers. In *Izvestiia*, the item was placed below another *TASS* piece about UNICEF, which announced the USSR's desire to eliminate all nuclear weapons by the year 2000. Later in the evening *Radio Moscow* stated that in addition to the two deaths, 197 persons had been hospitalized, of whom 49 were released after a check-up. The radio added assuringly that the "radiation situation" at the plant and in the adjoining area was "improving" and that the quality of the drinking water and the water in rivers and reservoirs was "in line with standards."

Soviet television released a photograph of the Chernobyl plant, which, it was claimed, had been taken shortly after the accident and revealed that the plant was not in ruins. *Radio Kiev* began the first of several broadcasts about the accident with untypical carelessness in a statement that "only" two people had been killed—the "only" was removed from subsequent broadcasts. The station was concerned to quell "western rumours" of thousands of deaths as a result of the accident.

1 May 1986

The overwhelming impression one gets from these first Soviet reports about a nuclear disaster is that the leaders were anxious to demonstrate that the situation was under control. The reports responded first to Western stories and only second to the accident itself, which may have been habitual but indicates a basic insecurity within the Soviet leadership. The First of May marked the start of a long weekend in the USSR, beginning with the May-Day celebration and followed with a bicycle race that was to commence in Kiev and continue through Eastern Europe, ending in Prague. By all accounts, the nuclear disaster 150 kilometres north of Kiev, which thus far had remained unexplained in the Soviet press, radio and television services, made not the slightest difference to the May-Day holiday events.

On 1 May, almost completely obscured by details of the celebrations, *Izvestiia* included the 29 April statement from the USSR Council of Ministers about the Shcherbyna commission and the alleged two deaths. *Radio Kiev* announced that celebrations would begin at 10 am and that tens of thousands of residents of the Ukrainian capital were participating, including guests. The Ukrainian party and government leaders turned out in force, confirming impressions that all was normal in the Ukrainian capital: First Party Secretary Volodymyr Shcherbytsky was accompanied by Second Secretary O.O.Tytarenko and other party and government leaders.

Nevertheless, the Soviet government was informing these same leaders and those celebrating (via *Radio Kiev*) that a further 18 people were in grave condition as a result of the disaster. The report declared that radiation levels had fallen by 1.5–2 times without giving any indication of how high they had been in the first place. The overall impression was of a surrealistic charade: a terrifying accident had occurred north of the city, the full effects of which were not yet known, and youngsters were dancing, singing and carrying flowers in the streets of sunny Kiev under the gaze of Ukrainian leaders, seemingly oblivious to the obvious dangers. Rarely has a regime displayed so knowingly an attitude of either disregard for danger or ignorance of the true nature and magnitude of the event that had occurred.

May the First saw Soviet officials outside the country making a series of statements about Chernobyl. These began with Iurii Dubinin, the USSR Ambassador to the United Nations, who addressed a meeting of the session in New York City. He asserted that the situation was very much in hand. "There is no need," he said, "for assistance from other countries." Soviet Foreign Ministry spokesman Vladimir Lomeiko was even more sparing in his comments. Denouncing what he called a Western "campaign" against the USSR "that does not want to acknowledge the data the Soviet government is providing," he informed viewers of *ABC Television* that the drinking water around Chernobyl and around the central plant "is very good and safe for drinking," a statement that was later contradicted by his own government.

In Washington, the Second Secretary of the Soviet Embassy Vitalii Churkin made an appearance before a House of Representatives Sub-Committee investigating the accident, but like his compatriots, he revealed nothing new, although he did state that the accident "was not over with." The most remarkable thing was perhaps that Churkin appeared at all, which does indicate that the Soviet authorities were prepared to uncover more to the U.S. government than they had revealed to their own citizens thus far.

2 May 1986

On Friday, 2 May, Boris Ieltsin, Candidate Member of the CPSU Politburo, who is also chief of the Moscow City party organization was in Hamburg to address the Thirteenth Congress of the German Communist Party (KPD). Focusing on what he called "our responsibility for the survival of mankind," he declared that:

Our ideological opponents do not miss a single opportunity to launch yet one more campaign against the USSR....The bourgeois propaganda media are concocting many hoaxes around the accident at the Chernobyl atomic power plant. And the purpose of all that is to step up even more the anti-Soviet hysteria in the hope of driving a wedge in the Soviet Union's relations with other countries. I can state with responsibility that the government is doing everything to eliminate the consequences of the breakdown, and, in implementing the energy programme, to continue using the atom for the peaceful purposes of the interests of man.

Ieltsin did not reveal at this stage, however, what the consequences of the accident were and or what might have caused the accident in the first place.

TASS turned its attention to Britain, in particular to the accident at the nuclear power plant on the south coast of Cumbria at Sellafield (formerly Windscale) at which an accident had occurred on 11 November 1957. Quoting *The Guardian*, *TASS* said that 13 people had died as a result of a "major accident," while "260 people were doomed to suffer from serious ailments caused by radioactive contamination." Later in the statement, *TASS* stated that since that time nuclear alarms had gone off at Sellafield almost 300 times, including four times in the first three months of 1986.

One can only assume that after Three Mile Island, it was considered expedient to focus on Windscale as the second (admitted) most serious accident to have occurred hitherto. In its critique, however, *TASS* may not have done a service to the Soviet cause. While Ieltsin was speaking of the uses of the "peaceful atom," the *TASS* statement was providing useful ammunition to environmentalists and the anti-nuclear lobby, groups regarded traditionally with some disdain by the Soviet leadership, but quite influential in countries such as Poland and Hungary. At the same time, five days had passed since the first Soviet statement on Chernobyl, and as it transpired, eight days since the actual event, and the Soviets were still hesitant to release facts.

3 May 1986

On 3 May, *TASS* revealed that a delegation of senior CPSU officials had visited the accident area on the previous day. The delegation included Egor Ligachev, a member of the Politburo and a Secretary of the CPSU Central Committee, Nikolai Ryzhkov, Politburo member and Soviet Prime Minister, and Shcherbytsky, the Ukrainian First Party Secretary. Other prominent officials present included Boris Shcherbyna, Ukrainian Prime Minister Oleksander Liashko and Hryhorii Revenko, a fairly recent appointee to the position of First Secretary of the Kiev oblast party committee. Evidently they visited the areas in which evacuated families had been located, making inquiries about medical and employ-

ment facilities, and about the operations of schools and preschool institutions. *TASS* noted that the delegation "decided on additional measures to
deal with the effects of the accident," implying that the steps taken initially had not been adequate.

The seniority of the leaders of this delegation indicates the import of
the accident. But at the same time, the evacuees were taken to neighbouring counties, to villages located at least 50 kilometres from Chernobyl. It
was not clear whether Ligachev and Ryzhkov visited the accident scene
itself. On this same day, Ieltsin made some further statements, one of
which was to reveal that on 2 May, radiation in the area around the
Chernobyl complex had been "under 200 roentgens per hour" falling to
100 by the following day. Radiation levels were thus admitted to be
dangerously high in the vicinity of the damaged reactor.

4 May 1986

Soviet viewers were given their first detailed look at the effects of the
accident on the *Vremia* newscast at 1900 hours, which provided shots of
the damaged installation as taken by a helicopter pilot, who was filming
without a mask of any kind. The newscast said that:

> As you can see, there is no vast destruction about which the ranks of the
> Western mass media have not stopped talking. Only the power set is
> damaged. All the production sites and neighbouring buildings as well as
> the supports for electric power lines are intact. Special units equipped with
> modern and effective equipment are carrying out work to clean the polluted
> areas adjacent to the territory of the station.

The newscast pictures revealed no people or animals in the area, however, and the only traffic in view consisted of a lone minibus. Moreover,
the newscast acknowledged that some areas were "polluted," which
negated earlier official comments about the purity of the water supplies.

The Ukrainian press gave front-page coverage to the visit of the Liga-
chev-Ryzhkov delegation to the nuclear plant, as reported by *TASS* the
preceding day.

5 May 1986

On 5 May, the world was still awaiting details about the scope of the
accident. Both *Pravda* and *Izvestiia* finally included longer articles on

the nuclear industry, but in neither case was the object to enlighten the reader about the events at Chernobyl. *Pravda* monitored what it called "attempts by Western countries to use the accident for political ends." Those who had offered aid to the Soviet government, however—countries, companies and individuals—were thanked. *Izvestiia* carried an article on an inside page entitled "Accidents at Nuclear Power Plants," which was divided into two columns: the top half was devoted to accidents in the United States; and the bottom half to incidents at plants in the United Kingdom. Chernobyl, which must have been the reason for the unusual attention to nuclear energy in the West, was not even mentioned.

Boris Ieltsin in Hamburg was more forthcoming. He informed the *Reuters* news agency that radiation was still being emitted from the damaged reactor, but maintained that further leaks had now almost been plugged. Helicopters were said to be dropping bags of sand and boron onto the reactor to plug the leak, while the authorities at the site had begun to "deactivate" the soil. *TASS* stated that work was under way to bank up the Prypiat River in the area of the nuclear plant "to prevent its [the river's] possible contamination." It also declared that the "radiation situation was stabilizing" in Ukraine and Belorussia, a comment that must have brought some relief to Belorussians. Their republic had been directly affected by the first cloud of radiation to be given off from the accident, but hitherto Belorussia had not been mentioned in the main Soviet reports.

The *Vremia* newscast carried an interview with Mikhail Krutov about life during and following the Chernobyl accident, which at first focused on a state farm located 40 kilometres north of the city of Kiev, i.e., only about 90 kilometres south of the Chernobyl plant. Radioactivity and monitoring stations had been established on the farm, but otherwise farmworkers were depicted going about their regular tasks. The programme then switched its attention to the streets of Kiev, which was shown making preparations for the bicycle race that was due to begin on the following day. Again, "normality" was the watchword, although some interviews with citizens were surprisingly revealing, given that the programme would have been edited before going on the air.

Krutov was shown stopping a passer-by and asking: "Excuse me please, can I ask you a question? Voices in the West are going on and on interminably about panic in Kiev, in your oblast here." The man responded candidly: "You know there is probably no panic. But we are worried about it too." Subsequently, Krutov interviewed a group of city bus drivers: "How was your first day back at work after the holiday?"

Well we all had to work over the holiday to evacuate people from the area around Prypiat and Chernobyl. We worked on transporting people out, on

11

evacuating them. In particular, we drove people from Radianskyi raion. The organization of the transport was, you could say, good.

Vremia indicated that first, a systematic campaign of evacuation occurred over the May-Day weekend, but second that the evacuation, or part of it, must have occurred several days after the accident took place. Even according to the Soviet accounts released by this date, which indicated an accident sometime before 28 April, this still meant that the evacuation was taking place a full three days afterward. And already Western reports were surmising that the accident had occurred over the *previous* weekend. The newscast demonstrated that there was no pre-planned local procedure for evacuating the plant in the case of an accident. A giant structure, at which a fifth reactor was almost ready for operation (as will be discussed below), was erected without any preparation for a worst-case scenario. There was no transport service in Prypiat capable of effecting an emergency evacuation of its citizens.

In one area, nonetheless, the Soviets had taken steps unannounced. The state-run Polish television announced on 5 May that Soviet nuclear experts had been in Poland "since last week" to consult with Polish specialists about preventative measures following the Chernobyl accident. The Soviet delegation reportedly had held talks with Polish Deputy Premier Zbigniew Szalada, the head of the Polish Government Commission set up in response to the disaster, and with Environmental Protection Minister, Stefan Jarzebski. The leader of the Soviet group was Valentin Sokolovsky, a Deputy Chairman of the USSR State Committee for Hydrometeorology and Environmental Control. It is not unlikely therefore that the Polish statements of 29 April about iodine preparations and other precautionary measures, which appeared so much more humane than the political statements about nuclear power in the West issued by *TASS* and *Radio Moscow*, were released after discussions with Soviet officials. Why then did the Soviet government not announce (or publicize) similar precautionary procedures for its own citizens? One can only conclude that the government knew of the dangers on 29 April and yet did not do anything.

6 May 1986

On 6 May, *Pravda* finally published a detailed account of the accident, although the precise technical details about how it occurred and how much radiation was given off were omitted (as far as the former was concerned, the probability is that no information was available). In an article entitled "The Station and the Surrounding Area: Our Special Correspondents Report from the Region of Chernobyl AES [Atomic Energy Station]," V. Gubarev and M. Odinets began by looking at the deserted city of Prypiat:

> Prypiat looks strange and unusual from the helicopter. Snow-white multi-storey buildings, broad avenues, parks, stadiums and playgrounds alongside kindergartens and stores....Just a few days ago, 25,000 power-workers, building workers, chemical industry and river workers lived and worked there. But now the city is empty. Not a single person on the streets, and no lights in the windows at night. And only occasionally does a special truck appear on the streets—the radiation monitoring service....Specialists are monitoring the station's reactors, which have now been shut down.

The reporters then provided the first account of the mishap itself:

> An explosion blew the roof off the [fourth] reactor. Structures collapsed over it, and a fire broke out. This happened at night. At the alarm signal from the fourth power unit, lieutenants V. Pravik and V. Kibenok, chiefs of the AES fire crews, quickly roused their firefighters. After the explosion, the roof of the machine hall had caught fire, and they focussed their efforts on putting out the fire. They fought the fire at a height of 30 metres. The firefighters' boots stuck in the bitumen melted by the high temperature and it was difficult to breathe because of the smoke and heat.

Noting that the firefighters' actions considerably reduced the potential damage, the reporters claimed that the accident was an event that had always been feared by physicists:

> The reactor's armour-plated core was exposed, some radioactivity was released upward and then a fire began inside. Further, it was particularly difficult to extinguish it because neither water nor chemical means could be used—the high temperature would instantly vaporize them and send them into the atmosphere.

While there was declared to have been no panic, "there were some scaremongers" among the thousands who lived near the plant. But when had the accident occurred? Without being specific, Gubarev and Odinets provided the answer when they noted that Z.F. Kordyk, the chief of the Chernobyl meteorological station, located at the juncture of the Uzh and Prypiat Rivers—presumably, then, within the city limits of Chernobyl itself—perceived increased radioactivity on the instrument readings "early that Saturday morning." At the same time, upon learning of the accident, bus drivers in Kiev "volunteered their services, even though it was a Saturday." In other words, the accident had occurred either very late on the Friday night of 25 April, or in the earliest hours of Saturday, 26 April (as was later revealed by Shcherbyna). In either case, the time was earlier than some Western sources had speculated.

According to the *Pravda* account, the evacuation of the immediate area was carried out in less than four hours, but this timespan did not include the time it would have taken for the drivers to complete the 130–150 kilometre journey from the city of Kiev to the plant, which could hardly entailed less than two hours, and given the notorious condition of Soviet highways, may have taken three. The evacuees were then transported to neighbouring raions—Ivankiv and Borodianskyi—where "domestic, trade and medical services" were provided for them. One case was cited in which a resident of Blidcha, Ivankiv raion took in 10 people from Chernobyl. Blidcha, however, is located only about 60 kilometres from Chernobyl and has no hospital according to the most recent Soviet account. It does however, possess a sizeable collective farm, and the evacuees seem to have been directed to villages at which they could be put to work on the farms.

Those evacuated to villages like Blidcha were clearly not in need of urgent medical attention, but this does not mean that they were unaffected by radiation leaks. *Pravda* said that Kiev's doctors "responded solicitously to the misfortune" and went into their hospitals. Three doctors from Kiev's "October" hospital established a medical section "at the site of the calamity." Others evidently came to Chernobyl from the "25th hospital." Thus within a few hours of the accident, some doctors from Kiev hospitals were on hand. How many were at Prypiat and how many at Chernobyl was unclear. Nor was it clear whether Chernobyl had been evacuated.

The *Pravda* article also said that technicians remained at reactors one, two and three of the station, which were still in the cooling-down process, having been shut off after the accident at number four. The article gave a fuller but far from comprehensive account of the events of 25–26 April and the aftermath of the accident.

On this same day, *Radio Kiev*'s domestic service stated ominously that because of wind shifts over the previous few days, an increase in radioactive contamination had been observed both in the city and in the district of Kiev. An official of the Ukrainian Ministry of Health, Anatolii Romanets, advised citizens to stay indoors as much as possible, to clean clothes from dust after returning home, and that people should eat food enriched with vitamins, especially Vitamin B. On the other hand, they were warned to avoid leafy vegetables such as spinach, sorrel and salad.

Two other events of 6 May illuminated the Chernobyl accident further.

Borys Shcherbyna, head of the USSR Government Commission set up to investigate the accident, held a news conference in Moscow, to which foreign journalists were invited. Ostensibly this conference was in line with CPSU General Secretary Gorbachev's policy of more openness in reporting events, a policy that had been cited pointedly in *Izvestiia* of the previous day. As part of this new "frankness," Shcherbyna declared that after the accident, local authorities had at first "underestimated its scope" and that evacuation had begun only the next day. As he gave the date and time of the accident as 26 April at 1.23 am, his statement plainly contradicted the *Vremia* account that had surely been viewed by the majority of those present at the news conference. For how could the drivers have worked on their free Saturdays in evacuating people if the entire evacuation process began only on the Sunday?

The "four hours" required for the main evacuation of personnel now took on a new meaning because the implication from Shcherbyna's statement was that this was four hours on the Sunday. He stated that over 100 people suffering from radiation were brought to *Moscow* on the night of 27 April, which at the earliest ("night" could hardly have referred to a time period before 5 pm) must have been a full forty hours after the accident occurred. Altogether, he stated, 204 people had been hospitalized, of whom 18 were said to be in a "serious condition."

More information was forthcoming from an afternoon programme of *Radio Moscow*, in which Vladimir Sokolov reported from Chernobyl, where a skeleton staff of 150 people were still at the nuclear power plant. As far as the evacuation was concerned, said Sokolov, about 1100 buses had carried out the process at the city of Prypiat. Presumably each would have only had to make one trip during the evacuation, given that the population of Prypiat was reported somewhere between 25,000 and 40,000 at the time of the accident.

Further questions remained. As the neighbouring communities consisted of farming personnel, and as at least two collective farms were in the area to be evacuated, what provision was made for the animals during the process? Were they simply abandoned to be slaughtered later? It is known that dairy farming predominates in the Chernobyl raion of Kiev oblast because the land is on the whole too swampy for grain farming, so there could have been a substantial quantity of cattle on the collective farm (the republican average is about 1,900 per farm).

Shcherbyna's press conference not only raised many more questions than it answered, it cast doubt on the alleged smoothness of the evacuation operation. Contradictions were appearing in Soviet accounts that were all the more disturbing given the time lag before the release of any information about the accident. If a Kiev bus driver had worked on Saturday, 26 April, as stated, then one must assume that he drove his vehicle to Prypiat and remained there for over 24 hours before the evacuation began on the Sunday afternoon. This seems improbable, because if he drove northward in the first place, he must have received orders to do so, in which case a high-level decision to evacuate had already been made. In short, official statements were not providing Soviet citizens with the truth on 5–6 May, and the likelihood is that the main culprit was the Moscow *Vremia* newscast hosted by Krutov.

In the eyes of many Western observers, the Soviet authorities were now at fault on four counts. First, they had delayed reporting an event of world dimensions that was a health hazard to both Soviet and non-Soviet citizens. Second, they had delayed the evacuation process until about forty hours after the accident occurred, which may have been fatal to some of those affected by burns or radiation poisoning. Third, they had not been truthful in some of their reporting, and official statements had begun to contradict one another. Fourth, the *TASS* news agency and *Radio Moscow* in particular had kept up a barrage of anti-American propaganda from the first days of the event, especially about the "thousands" of accidents to have occurred at U.S. nuclear power plants. To those who anxiously awaited news of the event, perhaps most particularly to those of Ukrainian descent in the West, this sort of reporting was extremely insensitive.

Perhaps because *TASS* had been given a propaganda mission to fulfill and was almost totally preoccupied with digging up accounts of problems at Western nuclear plants, it was left to the second Soviet news agency, *Novosti*, to provide the first grim depiction of the scene at the Chernobyl plant and in the surrounding villages. In a 6 May broadcast, correspondent Vladimir Kolinko's account of Chernobyl two days after the accident was released. He claimed that on Monday, 28 April, he had telephoned a former classmate of the Kiev Polytechnical Institute who was now a leading member of the Ukrainian Ministry of Power and Electrification. The classmate informed him that one man had been killed in the accident at the fourth reactor, while a second was missing, probably crushed when concrete slabs collapsed. "Several dozen personnel" had been badly contaminated.

On Tuesday, 29 April, Kolinko drove along the highway from Kiev to Chernobyl. When he arrived at Ivankiv [population 6,400—80 kilometres from Kiev], the fourth largest town in the area after Prypiat, Chernobyl and Poliske, he encountered "columns of tarpaulin-covered trucks"—i.e., there were large numbers of army personnel within 50–60 kilometres of the accident area. In Chernobyl itself, said Kolinko, "there were too many cars with Kiev number-plates in the town centre." [Party officials? Doctors and medical personnel? Nuclear experts?] In Chernobyl, he sought Borys Shcherbyna, but the head of the government commission had gone on to Prypiat, 20 kilometres away. Kolinko made the short drive to Prypiat, where he found every room in the party head-quarters occupied by power engineers and physicists, clad in blue and green protective clothing. With a pass signed by the Prypiat mayor, the journalist was allowed to drive around the nuclear plant town.

I interviewed a traffic militia major at a crossroads crowded with blue and yellow militia Ladas. ''The locality is badly contaminated near Prypiat, be-tween the Cloverleaf entrance and the turn to the station. Don't stop there, close the windows and put on your mask,'' he instructed.

Army engineer units were lining the roads of nearby villages such as Kopachi [8 kilometres from Prypiat] and Lelev [adjoining Kopachi]. Helicopters stationed in fields took off periodically for the nuclear plant (it had been revealed earlier that the helicopters were depositing sand and boron on the burning reactor). Trucks of the mobile contamination post were visible, and Kolinko reports that dosimetrics [people who measure radiation levels] in protective suits checked all the traffic leaving Prypiat. At this stage (29 April) the emergency headquarters had been set up in the premises of the Prypiat party committee—it was soon removed to the slightly safer venue of Chernobyl. According to Kolinko, the ground floor of the building was full of crates with rubber protective suits and masks. He maintained that the headquarters had been established ''some hours after the accident,'' at which time ''it was still not clear whether the population would have to be evacuated.''

Party members were preparing for an evacuation procedure, neverthe-less, and their main concern was to prevent outbreaks of panic. Once the decision to evacuate was taken, 1,000 buses removed the ''forty-odd thousand'' people ''in less than an hour.'' Here Kolinko's account is ei-ther flippant or inaccurate. We have already heard that the evacuation took four hours, and not less than one, which would have been a logistic impossibility given the number of people involved.

Kolinko went on to describe how a helicopter flew over the plant it-self, from which ''a wisp of smoke was rising.'' Thus the plant was pos-sibly still on fire two days after the accident. As for the injured, Kolinko cited a conversation he had with a doctor in an overcrowded hotel room in Chernobyl, which he shared with five other people—Chernobyl evi-dently had not been evacuated, despite earlier Soviet accounts that seemed to indicate that this had been the case.

A doctor...told me about 150 people who had been put in hospital. Later I was to learn from a government statement that the number was 197. My room-mate must have meant *only those who needed prolonged treatment*.

About ninety people from among the station personnel had suffered badly from exposure. The others were from the fire-brigade which came to the spot right after the explosion, guards, doctors, and car drivers who had been the first to arrive at the place of the accident [here Kolinko is surely referring to party officials, and not private citizens]. My doctor room-mate told me that some people on the station staff had been victim to sheer curiosity. Due to work on the morning shift, they were at home during the explosion. Even after the railway viaduct was closed to traffic and pedestrians, some people made a detour to get to the contaminated site across the railway.

Kolinko thus provided some very basic and important information: an explosion had occurred at the nuclear plant, as a result of which 150 people had been seriously injured and two others killed outright; the plant was still on fire; an emergency headquarters had been established in Prypiat and all the staff there were wearing protective clothing because radiation levels were dangerous; dosimetrics were checking all traffic and warning drivers to keep their windows closed and not to stop; Chernobyl 20 kilometres away had not been evacuated and in fact was badly overcrowded with emergency personnel. Had this kind of information been released on the day it was written, it would have provided Soviet citizens with enough information to make an informed analysis of the situation, and it would have assuaged some of the Western curiosity that was turning into anger at the lack of information over a major nuclear catastrophe.

But Kolinko's account was only released on 6 and 7 May. For nine days, the authorities had kept it in storage awaiting the right moment for release. Why? Was it too frank? Would it have led to the panic that according to Kolinko the officials in Prypiat were so concerned to prevent? Nothing seems to have been censored in the final analysis, so one can only assume that the authorities simply did not want this sort of information to be disseminated on 28–29 April. But it gives the lie to the future claims on the part of the Soviet authorities that the news services did not release information initially because they were still analyzing a "complex situation." The report was there. But it was kept on hold.

While Kolinko's report, even after its delayed release, was the frankest statement to have emerged from the Soviet side, it remained an isolated instance. Far more attention was being devoted in the Soviet press and on Soviet radio stations to past accidents at plants in the West, to the alleged nuclear weapons build-up in the West, and to past "misdeeds" of Western governments.

On this same day, for example, the newspaper *Sovetskaia Rossiia* was writing that the U.S. nuclear tests of April 1986 (in apparent disregard for the moratorium on such tests imposed by Gorbachev in the USSR) and the "bandit raids of American pilots" on Libya had endangered the entire world, while upon hearing about Chernobyl, the "propaganda machine" of the West had embarked upon a "hysterical anti-Soviet campaign." In an evening broadcast, *Radio Moscow* cited a report from the Washington correspondent of *Izvestiia*, which stated that "imperialist circles" were trying to transform a technical accident into an international conflict. One thing that should be noted, however, was that at last, Chernobyl was a news item in the USSR, even though it had yet to make the front page of a major Moscow newspaper. Ten days after the fact, the authorities had responded.

Radio Kiev, for example, began the day with a 6 am broadcast that repeated the announcement of the USSR Council of Ministers from the previous day. The statement of the Ministry of Health official Romanets followed. At 1900 hours, the radio's international desk highlighted the Moscow conference at which Shcherbyna made his comments. *Radio Moscow* devoted at least four of its 6 May programmes to Chernobyl, although two of them reverted frequently to anti-American statements.

As 6 May drew to a close, Soviet citizens were better informed about some of the events of the USSR's first reported nuclear accident. But Western correspondents who had attended the Moscow news conference in full force, as well as the Soviet public, still had no answers to several very basic questions: how much radiation had been released from the accident? What was the current level around the plant and surrounding areas? Above all, why had there been the delay in reporting the accident and in alerting Soviet citizens to the danger that surrounded them?

7 May 1986

Wednesday, 7 May saw two *TASS* reports from the city of Kiev, which were to some extent conflicting. The first declared that Kiev was "living a calm, confident and full-blooded life," whereas the second, highlighted below, indicated for the first time that there was an atmosphere of panic in the city:

The radiation situation which has arisen in Kiev at the present time does not require the application by the population of medical prophylactic measures. Furthermore, the unsupervised taking of various medicines, so-called "self-treatment," may be detrimental to health....Some of the city's hospitals for infectious diseases have indeed reported to us some instances related to the events at Chernobyl. Life is life, and there are always panicmongers. Heeding ill-considered advice, some people have taken medicines which supposedly give protection against radiation. They had the opposite effect—the result was poisoning. Now they are being treated for this.

In short, Kiev citizens were suffering because of a dearth of government advice—unlike their counterparts in Polish cities, who at least had some rudimentary guidelines to follow. Evidently citizens were administering iodine to themselves with catastrophic results. *TASS'* dismissal of such people as "panicmongers" was neither fair nor accurate. People were realizing that the situation in the city was worse than had initially been painted by the government, but were uncertain over what action to take.

One action the government of Ukraine was obliged to take was the provision of additional transport out of the city. As *TASS* stated:

> Of course, there is also disquiet. Especially among parents with regard to their children. The summer holidays are approaching. Lines for tickets have appeared at railway and Aeroflot ticket offices. Tens of additional long-distance trains, and flights by Aeroflot which were not scheduled originally are now being allocated. All the children from the evacuated raions will be the first to be sent to pioneer camps, to sanatoria and to rest homes.

In addition to the evacuees, ordinary city residents were now joining a growing number of people anxious to leave Kiev.

8 May 1986

In the period described above, which was one of the most confusing for Soviet citizens for many years, only the USSR Council of Ministers (the Soviet government) had made any official announcements. The more important body, the Communist Party of the Soviet Union (CPSU), as represented by the Politburo, had made not a single statement about the nuclear disaster. The official silence was reminiscent of Stalin's silence in June 1941 after the German invasion of the USSR. The only indication of the Politburo's concern had been the arrival in the area of the disaster of Egor Ligachev, a senior Politburo member and Party Secretary, on 2 May. And after his visit, Ligachev had left the statements to the Kiev First Party Secretary, Revenko.

On 8 May, the CPSU Politburo broke its long silence with the publication of a decree, "Concerning payment and the provision of material benefits to the workers of enterprises and organizations in the zone of the Chernobyl atomic power station." The decree examined the job placement of the thousands of people evacuated from the plant area, reparations for "damages caused to them" and rates of pay for jobs involved with "the removal of the consequences of the accident." It noted that those working in the area of the station—the contaminated zone—were receiving "additional incentive funds," while money was allocated in lump sums to the evacuees for meals and travel expenses. Disability benefits in connection with the accident were to be paid at the rate of average earnings "regardless of the work record or job."

The first party statement, then, was concerned with the basic living and working conditions of those linked to the plant. "Danger money" and "compensation" were the two watchwords. The former was probably of little solace to those who were working to put out the graphite fire at the fourth reactor. Two other reports from the Soviet press demonstrate that conditions now were especially hazardous at the nuclear plant.

In *Pravda*, Soviet Academician Evgenii Velikhov wrote about what he called an "unusually difficult situation" at the nuclear station. The accident, he declared, had led to a situation which required solutions that had never been faced before by scientists or specialists. An "offensive" had been mounted against the reactor, and workers were struggling "not only near it, but also underneath it. The task is to neutralize it completely." Was this an allusion to the threatened meltdown, to the danger that the fire might burn through the bottom of the plant and into the earth itself?

Writing in the Ukrainian workers' newspaper, *Robitnycha hazeta*, three *TASS* correspondents V. Zhukovsky, V. Itkin and L. Chernenko revealed that there had been a danger that the fire might spread from the fourth to the third reactor. After the "first victory" over the blaze (to keep it under control), the toll had been heavy: seventeen firefighters were hospitalized. In addition to the Soviet Commission under Shcherbyna, the firefighters and militia in the entire area were subject to the jurisdiction of the Ukrainian Minister of Internal Affairs, I.D. Hladush. According to the newspaper account, a full-scale Ukrainian military operation was now combatting the fire at the atomic plant, while the official government commission under Shcherbyna made its investigation.

By the end of Thursday, 8 May, the full extent of the disaster at Chernobyl was becoming more apparent through the official Soviet news services. The Soviet government and the CPSU were starting to behave in what one might call a "normal manner" for an emergency of such proportions. At the same time, this was not an admission that the previous silence over the accident had been a mistake. In fact, the more action the Soviet authorities took in connection with the disaster, the more virulent

21

were the attacks on the Western reporting about the event. The official Soviet view was that the Western media and Western governments were making political capital out of a life-threatening situation, and that the potential death toll was the cause of great glee in Western circles. Whenever possible, the Soviets used statements by East European government news services (especially those of Czechoslovakia) and, especially, by Western visitors in the USSR to condemn those behind the "anti-Soviet campaign."

> [Odessa.] A large group of priests and parishioners of the U.S. Episcopal Church which is in our country to attend the Orthodox Church's Easter celebrations has condemned the hostile anti-Soviet campaign launched in the West in connection with the accident at the Chernobyl AES. In an interview with a TASS correspondent, James S. McReynolds, leader of the group, said that he had a high opinion of the Soviet people's love of peace...and he supported the Soviet initiatives aimed at preventing a thermonuclear catastrophe. [McReynolds said] we condemn the campaign of fabrications launched against the Soviet Union in a number of Western countries, including the United States, over the incident. During our stay in the USSR, we have felt quite comfortable, nothing has endangered our health, and we have not seen any panic here.

The problem for the Soviets was that for every McReynolds statement, there were ten others suggesting that the dangers were very real. And most of these were now coming from the Soviet side.

On a *Radio Kiev* broadcast in the early evening, for example, Anatolii Romanenko, Minister of Health of the Ukrainian SSR, stated that "in the last few days," over 20,000 residents of the city of Kiev had been examined for radiation effects (although none were reported to be suffering). According to Romanenko, the main enemy was dust, which was a principal carrier of radioactive substances. In areas that had carried out extensive washing of streets, courtyards and squares, "the background radiation is several times lower [than in other areas]."

Romanenko advised listeners to shower daily and to wash their hair. He revealed that children of Kiev were now being kept indoors, mainly to protect them from dust, and that the school year was to be ended within one week for those schools that had taken in evacuated children and for Grades One to Seven. A *TASS* account datelined Kiev referred to a "battle situation" in the city. Specialists from Moscow and Leningrad had arrived there for "consultations and to provide assistance" to local doctors. They were led by Academician L.Ilin, the Vice-President of the USSR Academy of Medical Sciences. About 50 per cent of those who had been hospitalized immediately after the accident—about 100 people

—remained there on 8 May, almost two weeks after the event.

This same *TASS* report also cited the First Deputy Minister of Health of the Ukrainian SSR, A.N. Zelinsky, as stating that over 1,300 doctors, nurses, laboratory technicians and radiation monitors, along with 240 ambulances were working "in the region of Chernobyl AES [atomic energy station]." In the village of Piskivka (Borodianskyi raion), 80 kilometres southwest of Chernobyl, which had received 2,000 evacuees, 12 teams of doctors were reportedly sent from other oblasts to assist the local hospital staff in providing medical care for the victims. Squads of medical persons were thus being assembled in various villages, particularly in those which like Piskivka, possessed a hospital.

More conflicting statements did little to assuage citizens' anxiety. While *Novosti* was stating that "the locality was threatened with fall-out within a ten-kilometre radius" of Chernobyl AES, *Radio Moscow* was telling world listeners that every person who had been in a 30-kilometre zone around the nuclear plant had undergone "special treatment." Further, in Kiev itself, noted the radio, radiation monitoring posts had now been established at every crossroads. Food and the air were the special concerns:

> Everything taken to stores is checked and vegetables supplied to cities from nearby areas are checked twice, in the field and in the stores before they are put on the counters. Highways are washed in the Kiev district all the time and special tank trucks have been made available for that purpose.

Also on 8 May, the first evidence arose that the evacuation procedure in Prypiat, Chernobyl and other areas was not a smooth, well-organized process, but rather something approaching a fiasco. It came as a result of an odd meeting between O. Liashko, the Chairman of the Ukrainian Council of Ministers, and a group of foreign reporters who were given an official tour of the city of Kiev. According to the accounts of this meeting given by the major Western news agencies—*Reuter*, *AP* and *UPI*— Liashko's statements appeared to refute those of Shcherbyna at the Moscow news conference two days earlier.

Whereas Shcherbyna had blamed "local officials" for the delay in beginning the evacuation from Prypiat, Liashko said that Moscow was informed of the seriousness of the disaster only on 28 April because "the measurements at first showed there was nothing to fear." He also said that the evacuation was carried out in three stages: the first, on 27 April in the afternoon, involved those people within 10 kilometres of the atomic plant (*Novosti*'s "danger zone"); the second occurred on 2 May, and expanded the zone to 30 kilometres; while the third occurred as late as 4 May and it is not clear from Liashko's statement when it had ended.

23

Altogether, said the Ukrainian Premier, 84,000 persons had been removed from the zone, and not 49,000 as announced in official statements. Liashko's statement confirmed what some had suspected: that people in the 10–30 kilometre zone—which included the raion's capital city of Chernobyl—were only moved away from the accident site one week after it occurred.

Western writers speculated that a dispute had occurred between Ukrainian and Moscow officials. Among the more extreme speculations it was suggested that the Ukrainian Party Secretary's job might be on the line because of a dispute over the responsibility for errors made in the evacuation decision and procedure. Here is an example from *The New York Times*:

> Both the spokesman in Moscow, Borys Y. Shcherbyna, and Sokolov in Washington seemed to be suggesting that the Russians' early failure to alert the world resulted from local officials' incorrect assessment of the accident and "human error" in judgment. That could well be true given the tendency of bureaucrats everywhere to cover their mistakes—and particularly because Gorbachev, in his modernization campaign, has been rough on lower-level officials whose performance he considers sub-par. On the other hand, this could be a pass-the-buck maneuver to cover up on Moscow's own mistakes. Either way, it may portend unpleasant consequences for such local officials as Vladimir Shcherbitsky, the Ukrainian party leader, a holdover from the regime of Leonid Brezhnev, three governments back.

While it is plausible that local officials underestimated the situation, Chernobyl was not a mishandled grain harvest, but a unique and unprecedented event that had taken the entire country by surprise. The significance of the dispute, if indeed it can be called a dispute, is that the Soviet authorities did not have a clear idea of what constituted the danger zone. As described above, the city of Kiev was declared to be quite safe at first, but within one week to ten days, children were being hustled indoors. Likewise in the accident zone. Ten kilometres *seemed* safe, but it soon became evident that it was not. Possibly the extension of the danger zone from ten to thirty kilometres on 2 May was a direct result of the arrival in the region of the Ligachev-Ryzhkov-Shcherbytsky group. Certainly the times coincided and we know that "additional measures" were taken. But they were taken by both Moscow and Ukrainian officials, nonetheless. If there was a dispute, it was not between Shcherbytsky and Ligachev (or Gorbachev), but rather between both the above and local *raion* officials: the Chernobyl raion party secretary and the Prypiat city party committee secretary. These latter bodies have in any case little scope for independent decision-making and can hardly be faulted for a reluctance to take broader actions.

Another problem now emerges with regard to the evacuations. If 84,000 people were involved, this figure is greater than the officially reported population of the entire Chernobyl raion. Other raions may have been affected too. But initially, Chernobyl residents were simply taken to the neighbouring raions. Were they subsequently *re-evacuated*? If not, those moved to some parts of Ivankiv raion, for example, would now be living in the newly defined danger zone. So a big drawback in the evacuation was that the families transported were not always moved far enough away from the accident site.

In the Prypiat zone itself, residents were evidently desperate to leave the city. The 8 May edition of *Sovetskaia Rossiia* reported that after the disaster, a delegation of agitated Prypiat residents appeared at the party offices in the city, a few of whom "tried to make trouble."

9 May 1986

The Friday morning edition of *Izvestiia* also alluded to difficulties in evacuation with a statement that:

> There is no point in denying that there were individual workers who failed to show sufficient firmness or readiness to take decisions in the emergency conditions.

Of more significance in this same report was a reference to the fact that those monitoring the amount of radiation within the plant evidently were not in touch with those monitoring radiation in the vicinity. *Izvestiia* declined to elaborate on the significance of this fact, but the danger is lucidly clear. Those working and living outside the plant's zone were *unaware* of the dangerous emissions of radiation into the atmosphere. According to the earlier report about the Chernobyl meteorological office, people in that city realized that something was wrong at the nuclear plant, but they were not aware of the extent of the danger. This lack of communication constituted a major safety hazard and may have directly cost lives. If people outside the Prypiat area were not aware of how much radiation was being released into the atmosphere, then how could local officials in Chernobyl or even Kiev make a rational decision about when and how many villages to evacuate?

This same newspaper also published an item about the head of the fire brigade, 35-year-old Leonid Teliatnikov, who was interviewed from a

Moscow hospital. Alerted by the controller of the nuclear plant's fire unit (located six kilometres from the station), although officially on leave, Teliatnikov had arrived at the nuclear station's fourth reactor on 26 April, and had recognized at once that the biggest potential danger was that the fire would spread through the cable leads to the third (twin) reactor. Evidently he only had fifteen men in the unit and called immediately for reinforcements. Flames were "raging...in at least five places." It appeared that the fire might spread to the machine room, "and destroy the system for the protection of the entire power station." For three hours Teliatnikov and other firefighters reportedly battled heat, melting asphalt and poisonous fumes, working at times at a height of 71 metres, where the fire was at its most intense. At least two people were involved in preventing the fire from spreading to the third reactor, which involved "an incredible effort." All the firemen were said to be well aware of the danger they faced and the extremely high levels of radiation had already been monitored by dosimetrics. The fact that Teliatnikov had been brought to a Moscow hospital suggested that he was among the most severely injured, but the article focused on his bravery rather than the injuries he might have sustained. According to his own statement, his condition on 9 May was still "normal."

The International Atomic Energy Agency (IAEA) representatives, who had been permitted to visit the disaster scene, issued statements on 9 May, which provided some new information about events, but not about the specific cause of the accident itself. Morris Rosen, the Director of the IAEA's Nuclear Safety Division, revealed that:

> On the 26th April at 1.23 am, explosions occurred in the fourth unit of the Chernobyl nuclear power station. At that time the reactor, which was undergoing a planned maintenance shutdown, was at seven per cent power level.
>
> As a result of the explosions, two people lost their lives, one from hot steam burns, the other of injuries from falling objects.
>
> So far there are only hypotheses regarding the specific reasons for the accident. Research and detailed analysis are under way. Records of data from the control room were recovered after the accident and are being studied.

Rosen thus indicated that ultimately the precise cause of the accident would be known and that the reactor was not at its normal capacity when the accident took place. The explosions resulted in the fire, which caused "extensive damage" to the reactor itself and to the reactor core, "resulting in radioactive releases beyond the nuclear power station area."

About 50 per cent of the radioactive releases consisted of iodine-131, reported Rosen, and these would be short-lived.

As for the evacuation, it had begun on 27 April, beginning with women and children, and up to 48,000 people were evacuated from locations within a 30-kilometre radius.

As a preventive measure, potassium iodine tablets were widely distributed inside as well as outside the 30-kilometre zone. Two hundred and four persons, including nuclear power station personnel and firefighters, were affected by radiation from first degree to fourth degree, 18 persons being in the fourth degree. All 204 persons were hospitalized in Moscow and treated medically. In some cases bone marrow transplants were performed.

Rosen's remark that iodine tablets had been issued at the outset had not been heard before. At the same time, his report, which was uncritical in tone, mildly rebuked the Soviet authorities because "no systematic data on radiation levels were made available."

One result of the IAEA's visit to the Chernobyl plant was the agreement by the Soviet authorities to establish a permanent station some 50 kilometres from the plant, "to keep control over the atmosphere." *Novosti* reported that the station would be taking

around-the-clock measurements, daily aggregate data on the ecological situation and the level of radioactivity in the region [which] will be delivered to the IAEA for the further circulation in the states concerned.

This decision shows that the IAEA was concerned with the Soviets' previous failure to alert neighbours about the radiation cloud that had resulted from Chernobyl.

In general, however, the Soviet authorities were still very concerned to demonstrate that the worst was over and that no real danger was posed from radiation almost two weeks after the accident had occurred. Revenko, the First Party Secretary of Kiev oblast, was cited in *Pravda* as stating that while precautions were being taken, "the region keeps working." Potatoes had been planted ahead of schedule, and milk yields had increased, while the streets of Kiev itself were "as always, swarming with people," and "factories, shops and markets are working just as efficiently as before."

Radio Moscow, in providing an interview from the Kiev Centre for the Study and Monitoring of the Environment, also hinted that there was no further threat from radiation. Having described the testing of water samples from the Dnieper, Prypiat and Desna rivers (without providing any results), the head of the Centre, Ia.N. Pimenenko, assured listeners that:

> In the most general terms one can say the following. The radiation in the atmosphere, which already presents no danger, is on a downward trend. The water is virtually pure. With regard to the soil we are conducting our usual observations.

The account sounded similar to others about the Soviet nuclear power industry presented before the Chernobyl accident took place.

While the world still did not know the full extent of the radioactive fallout, one source reported that by 9 May, the level in Kiev had fallen considerably. According to the Yugoslav news agency *Tanjug*, Kiev's level was 150 microroentgens [per hour] on the morning of 9 May, "while in the oblast it ranges between tolerance levels of 20 to 160 microroentgens." But in *Pravda Ukrainy*, the Ukrainian Health minister Anatolii Romanenko said that while radiation levels were falling, people should still be taking every precaution to keep potential radiation contamination to a minimum.

10 May 1986

On Saturday, *TASS* provided some data from the USSR State Committee for Hydrometeorology and Monitoring of the Natural Environment, which stated that the radiation level was 0.33 milliroentgen per hour at a distance of 60 kilometres from the station and 0.32 milliroentgens in Kiev, "which is absolutely safe to people's health." This level, however, was twice that reported by *Tanjug* on the previous day.

More information about the radiation in the atmosphere after the disaster was released by Radios *Prague* and *Moscow* on 10 May. *Radio Prague*, which had steadfastly and rigidly adhered to the Soviet line in all its statements about Chernobyl, quoted a Soviet official as stating that iodine-131 had escaped from the damaged Chernobyl reactor and that this isotope was dangerous because it could enter the human organism through food. The statement was given in an interview with Iurii Izrael, the head of the State Committee for Hydrometeorology and Environmental Control by *TASS* correspondents. While the essence of the interview was broadcast by both Moscow and Prague radios, only the latter included the reference to the iodine-131. Izrael declared that the "development of radioactive gases and substances" had lasted several days and was connected to the high temperatures in the zone of the reactor. He referred to what he termed "a slight increase of radioactivity levels in some cities of Ukraine and Belorussia," while maintaining that these increased levels "pose no threat to the health of the population."

On this same day, the Soviet authorities once again reverted to the line that the main danger was now over. In a broadcast on a morning programme, *Radio Kiev* declared that it had reliable information from specialists at Chernobyl that "the situation is stabilizing." Day-by-day, it stated, the temperature in the zone of the damaged reactor was falling, and the level of radiation had fallen—although it was still unsafe "for the health of the people." Within a brief time, the programme reported, radiation levels within a 30-kilometre radius of the damaged plant had been brought down, and the isotopes released into the atmosphere were of brief duration, including iodine-131. And yet, work on the damaged reactor was still continuing.

11 May 1986

In an afternoon broadcast on *Radio Moscow*'s domestic service, Soviet Deputy Prime Minister, Ivan Silaiev, revealed that "the main danger at the damaged Chernobyl nuclear plant is now over." But while Western claims that a "massive catastrophe" had reportedly been proven erroneous, work must not be allowed to slacken off because "a whole series of complicated problems" remained to be solved. E.P. Velikhov, the Vice-President of the USSR Academy of Sciences, who had, it transpired, a major role in drawing up strategy for countering the effects of the accident, said that Sunday [11 May] "was a turning point." The feared "major threat" had evaporated, but a variety of examinations still had to be made before the Chernobyl nuclear plant could resume operations again. Soviet Academician Valerii Legasov was quoted by *Radio Moscow* as saying that "We must not only restore the atomic energy station, but also revive the surrounding land and give its life force back to it."

Later in the day, correspondent Aleksander Krutov again delivered a report for the *Vremia* Newscast, the main theme of which was to reiterate that 11 May was very much the "last day" of the Chernobyl accident. Here is one excerpt.

Krutov: I heard Academician Velikhov say just now that a historic event occurred today:
Silaiev: Well, in the main he is right. We have come to the conclusion today that the basic, main danger has been eliminated. This is of course an

historic event. Today we can already work more calmly, insofar as what the world predicted, in particular the western bourgeois newspapers that shouted from the rooftops that a colossal catastrophe was about to happen—today this does not threaten us. We are today firmly convinced that the danger is passed.

According to Silaiev, those working at the plant site had "taken a quick breath" and had begun to concentrate on issues that had previously been considered as secondary to the task of averting a major disaster such as a fire at the third reactor: decontamination, removal of the "remnants," diagnosis and monitoring, "and in the final analysis, we have to resolve the problem of burying this zone."

Silaiev's comments appeared to be reassuring but this was very much the intention as was demonstrated later in the programme when neither he nor Krutov could decide which of the photographs they possessed was the most recent to be taken of the nuclear plant.

Krutov: These photographs are the latest received from the station, are they not? [Video shows Silaiev with folder of photographs]
Silaiev: Yes, these are the latest. No, they are not the latest, the latest photographs are much calmer. Where are they? [Searches through folder] Yes, let's have a look, yes, this is the latest photograph. As you can see, this shows a completely calm state.

This remarkable dialogue on a Soviet television programme highlights its chief concern: to show the viewer that the main danger at Chernobyl was now over.

Two announcements from the USSR Council of Ministers effectively concluded the whole episode after a fashion. *Izvestiia* included another "From the USSR Council of Ministers" statement on 11 May, which outlined the work on the damaged reactor and on the Prypiat River of 8 and 9 May, and declared that the state of the first three reactors at Chernobyl station was "normal." It also noted the visit to the area of the station of H. Blix, the General Director of the International Atomic Energy Agency. Later in the day, *TASS* gave another announcement from the Council, stating that measures had been put into practice to encase the reactor at the fourth power-generating set in concrete, and gave reassuring information that there had been no change in the (already safe) levels of radiation in Ukraine and Belorussia. Again, the impression is that the crisis had been surmounted.

In many respects, this was a simplistic and transparent approach, particularly in view of the long-term effects of radiation. In fact, 11 May might have marked the beginning of the Chernobyl affair. The explosion, fire and ensuing evacuation had ended; but the questions, criticisms and recriminations were just beginning.

12 May 1986

Pravda finally began to apportion some blame for the delays in evacuating residents in the danger zone. In the report by Gubarev and Odinets, "Raikom Working Around the Clock," party officials with the "Iuzhatomenergostroitrans" (Southern Atomic Energy Construction Transport—or trucking agency) association were chastized for allegedly treating the evacuees with a callous indifference:

> It is 10 days since the collective at the Chernobyl subsidiary of the "Iuzhatomenergostroitrans" transportation production association, comprising more than 200 people, was evacuated along with the families, to Polesskyi and Ivankovskyi raions. However, during this time, the collective's leaders—Communists A. Sichkarenko and A. Shapoval—have essentially done nothing to help the people under their jurisdiction or to provide them with work. Wages have not been paid on time, clothes have not been allocated, and evacuees' legitimate requests have been ignored. Finding himself in Polesskyi raion among subordinates evacuated from Prypiat, A. Shapoval, the subsidiary's chief engineer, was entirely indifferent to the fate of people who found themselves in a difficult position.

As a result of these deficiencies, Shapoval was expelled from the party, while Sichkarenko was strictly reprimanded and had his party card endorsed by the Prypiat city committee session. Both were removed from their posts. *Pravda* observed that the unusual circumstances of the Chernobyl accident had "highlighted bottlenecks" and demonstrated that certain leaders had been "psychologically unprepared" to cope with such conditions. This statement contradicted earlier Soviet assurances about the model way in which the entire evacuation had been conducted. It was yet another event concerning which the real situation only emerged with the passage of time.

13 May 1986

Radio Moscow's "Iunost" programme of 13 May offered more insight into the immediate post-accident events. Correspondents reportedly met with "dozens" of witnesses, including the Deputy Minister of Internal Affairs of the Ukrainian SSR, Major-General Berdov. The latter was said to have been at the scene within ninety minutes of the accident. Members of the Government Commission arrived "several hours later" so that both Ukrainian militia and civilians in charge of the investigation must have been at the accident site before daybreak on Saturday, 26 April, a fact that makes the delayed evacuation all the more inexplicable. Certainly Berdov (or his superior, Hladush) made an immediate decision to evacuate people in the area at this time and "thousands of militia functionaries began this most difficult operation."

The programme also gave a succinct account of the sealing off of the damaged reactor from the air:

> It became necessary to close off the source of danger, to block it off, deluge it and seal it off. This could only be done from the air—and the aviators' turn came. Courageous helicopter crews carried out hundreds of flights and, in incredibly difficult conditions, the core of the fourth unit was sealed off by means of an enormous stopper, composed of sand and other materials, weighing in excess of 5,000 metric tonnes....Military helicopters overfly the station several times a day even now.

At the same time, a tunnel was being dug underneath the reactor evidently with the assistance of Kiev metro-builders, two hundred of whom were now working at the Chernobyl plant "to cool and strengthen the foundation of the reactor," according to a *Radio Moscow* report (1700 hours).

14 May 1986

On 14 May, Mikhail Gorbachev, CPSU General Secretary, finally made a television address about the Chernobyl disaster, more than half of which was devoted to the reporting about the accident by the West and the "mountain of lies" that had accrued. As for Soviet workers, Gorbachev praised their part in overcoming a stern test that had taken the lives of V.N. Shashenok, an adjustor of automatic systems and V.I. Khodemchuk, an operator, and led to radiation poisoning for 299 people, seven of whom had already died. While commenting that all nations involved in the production of nuclear power should co-operate with the IAEA, particularly with early warnings about radiation leaks, Gorbachev was careful not to apply any criticism to the way in which the disaster was handled by the Soviet authorities. He extended the USSR's moratorium on nuclear tests to 6 August, the forty-first anniversary of Hiroshima, but made no promises to consider postponing or altering the USSR's nuclear energy programme. The speech, then, was basically political in tone, and despite the condolences for those suffering from the effects of the accident, it showed little of the frankness and blunt commentary that had characterized some of the Soviet leader's earlier offerings. This reflects the sensitive nature of the subject and Gorbachev was apparently unwilling to say anything that might appear to compromise the nuclear power build-up in the USSR and Eastern Europe, which, as will be shown below, is perceived as of the utmost importance to the future of the Soviet Union.

* * *

This survey of the first days after the Chernobyl disaster as seen through Soviet eyes (and the full impact of the accident will be discussed later) reveals that it took the Soviet authorities some time to respond to the catastrophe. Only by 8–9 May had full emergency measures been undertaken and by this date, precautionary actions such as the banning of street vendors in Kiev, stringent norms imposed on all agricultural products and dosimetric check-ups throughout the city, were at last in effect. Those involved in international assistance to the victims of Chernobyl, such as the Los Angeles bone marrow specialist, Dr. Robert Gale, were apparently satisfied with the authorities' efforts at this point: the government was doing its best in a very complex situation.

None of this is in dispute. What is inexplicable is the failure of the Soviets to report the event or to take any kind of action in the first hours, and even the first days after the accident occurred. This raises some questions not so much about the morality of the government's stance—the main focus of Western reports—but rather about safety standards and the government's capability in the face of an emergency in the nuclear power industry. One would have expected the Soviet authorities to have been reluctant to divulge information, initially. This is ingrained in the system, habitual. But one would have anticipated that in an industry which is currently the subject of a massive expansion and build-up that the possibility of an accident would have been foreseen and taken into account. Why was there insufficient transport at Prypiat, or at least Chernobyl, for example? Why were two reactors built so closely together, so that an accident at one endangered the other? Why was the Chernobyl fourth reactor inadequately contained, as was clearly the case?

In addition to industrial safety drawbacks, one has to account for the apparent disregard for safety *after* the event. If ice cream and fruit vendors were banned from the streets of Kiev on 9 May, as reported in

Izvestiia, if schools were closed down on 15 May and children were kept off the streets two weeks after the accident, why was this not the case during the period from (approximately) 26 April to 5 May? Why were the May-Day celebrations in Kiev and the opening leg of the Kiev-Prague bicycle race allowed to take place? The Soviet authorities were hardly in a position to predict the direction of the wind, even from hour to hour, and we know from the appearance of Soviet environmental leaders in Warsaw that they were fully aware of the existing dangers. By these actions, Soviet leaders put in jeopardy the health of over 2.4 million residents and vacationers in the city of Kiev, most of whom were not aware of the perils they faced, having heard only reassuring comments from Soviet newspapers, television and radio.

An attempt at answering the lack of information aspect was made by Georgii Arbatov, the engaging head of the Institute for the Study of the United States and Canada, who held a series of interviews on 8 May, by which time he had presumably been well primed by his superiors:

> *Novosti*: Many people in the Western mass media accuse the Soviet Union of providing belated information about the accident. What can you say on that score?
>
> *Arbatov*: The situation required thorough check-ups, only dependable information, based on hard facts and instrument data, could be made public. First of all, endangered people had to be rescued. That accounted for many complications in studying the accident cause right after the event. Besides, it isn't easy to see what brought about a situation far out of the ordinary. American specialists, for instance, are still investigating the Challenger tragedy.

This explanation was hardly satisfactory. On 6 May, *Pravda* had informed readers that on the day of the accident the Chernobyl meteorological station monitored high radiation levels at the nuclear plant. At this point, the authorities could have alerted both their own citizens and neighbouring countries. As for the "rescue mission" mentioned by Arbatov, we know that this was not mounted until 27 April (about forty hours after the event).

Could one then say that the initial crisis at Chernobyl was a result of an information embargo imposed by a closed totalitarian society? The answer is that this is only part of the truth. After all, even though the government did not immediately ascertain the cause of the chemical explosion and fire at the plant, it did eventually release a considerable amount of information. What one can say with certainty is that with Chernobyl, the Soviet government had a lot at stake. Not only were lives imperilled, but an entire economic programme for the future was directly threatened by the accident, one that had only recently been announced with enthusiasm at the Twenty-Seventh Party Congress of the CC CPSU two months earlier. Nuclear energy is one of the key expansion areas of the Soviet power industry. The centre of this industry is Soviet Ukraine: Chernobyl happened at the worst possible time and in the worst possible region as far as Soviet leaders were concerned.

The nature of this development will be discussed below. Here it suffices to say that Chernobyl cannot be analyzed adequately without a knowledge of the recent developments in the Soviet nuclear energy industry. The significance of this sector goes a long way to explain the various contradictions in Soviet statements following the accident, and the obvious reluctance to say anything that might prejudice the future of the industry. To their own citizens and to the world at large, the Soviet authorities tried to convey the impression in the first days after the disaster that the situation was under control. But as the disaster grew in dimensions, the Soviet leaders eventually were obliged to take the precautionary actions that were either ignored or rejected earlier.

Was this an assault on the Soviet citizen? Or on Ukraine and Ukrainians specifically? Ultimately, no. It illustrated the priorities of a bureaucratic state; that to the Soviets, the economy of the country took precedence *in every situation*. But it can be argued that a healthy economy is essential to the future of the Soviet people, i.e., the needs of those in the danger zone of a nuclear accident are less important than the requirements of the entire Soviet population. Only by comprehending such an attitude can the reader begin to understand the aftermath of Chernobyl and the release of vital information in such a painstaking fashion.

Chernobyl was a symbol of the nuclear industry; by 1988, it would have been the largest nuclear plant in the USSR, and it was visible proof that the industry could provide a short-cut to success, a means for a major technological advance in the latter part of the century, away from the traditional Soviet reliance on fossil fuels into the nuclear era. Its beginnings lie in the Soviet energy question: difficulties in obtaining raw materials for the USSR's power industry, and a shortage of these same materials in many East European countries.

CHAPTER TWO

Soviet Energy in the 1980s

Soviet Ukraine, the location of the Chernobyl plant, plays an integral role in the USSR's Energy Programme, the outlook of which has been forecast up to the year 2000. It may seem paradoxical to refer to an "energy crisis" in the USSR, given that it is one of the few countries in the world that to date has not had to import energy resources. The difficulties that have arisen in this sphere have been a result less of the amount of raw material supplies than of the expenditures required to extract and transport them to the principal consuming areas in the European part of the Soviet Union.

Prior to the Twenty-Seventh Party Congress of the CPSU Central Committee in February 1986, the draft document on energy questions that was later to be approved with few changes by the Congress was analyzed by *TASS*. Having made reference to the country's plentiful natural resources, *TASS* declared on 7 January 1986, that:

> Scientists estimate that the USSR is not threatened with a shortage of raw materials; but the basic supplies of oil, gas, and coal are concentrated in remote and poorly developed regions of the east and the north, where there is a harsh climate, permafrost and no roads....Specialists consider that it is not profitable to expand production [of these resources] to the same major proportions as before—in the last decade, expenditure on the recovery of each metric tonne has increased by a factor of three.

The quotation was not strictly accurate, in that there are also plentiful coal supplies in the huge Donbass coalfield that stretches through Donetsk, Voroshilovhrad and Dnipropetrovsk oblasts of Ukraine into Rostov oblast of the Russian republic. This area was addressed by the

Minister of Power and Electrification of the Ukrainian SSR, V. M. Semeniuk, in November 1984:

> Coal remains the principal organic fuel. But the scale of its use is limited first and foremost by the complex conditions of its extraction.[1]

These difficulties will be discussed in more detail below. Suffice it to state here that of the three major energy resources in the Soviet Union: oil, coal and gas, only the latter has given cause for optimism over the past few years. And yet, gas reserves are to be preserved as one of the more reliable means of obtaining hard currency from the West in the future for the USSR. The extraction of gas is to be raised by almost one-third by the end of the century, but this is a relatively modest increase.

The Ukrainian SSR has long remained one of the main industrial regions of the Soviet Union, especially for its production of coal, steel and chemicals. Because of an abundance of natural resources, both Tsarist and Soviet governments have heavily exploited Ukraine's raw materials. An energy imbalance was created, whereby the Moscow government used the republic as a vital raw-material source, and exploited these resources almost recklessly, as the Soviet leaders have now conceded. Not only does Ukraine possess iron ore, coal and uranium in plentiful supply, it is also located close to the centre of the Soviet market. It is within easy reach of the USSR's major population centres and the Donbass-Dnieper zone is one of the major industrial regions of the USSR.

For the purposes of this study, the major problem in the Ukrainian energy sector is the Donetsk coalfield, traditionally the principal source of Ukrainian energy supplies. Stagnation in this coalfield is an important reason for the widespread development of nuclear energy in the Ukrainian SSR, as a replacement for fossil fuels. In theory, this problem should not have arisen because the Donetsk is not about to run out of coal. Yet for a number of reasons, both geological and man-induced, coal-mining no longer has a viable future in Ukraine. Nuclear energy has been handed its mantle.

Discovered in 1721, the Donetsk Basin has played the major role in Russia's (and later the USSR's) coal industry since the mid-nineteenth century. In 1913, it accounted for about 87 per cent of the Russian Empire's total coal production.[2] In the 1930s, however, the Soviet authorities began to develop coal resources in the eastern part of the country, where many of the deposits could be mined by the relatively inexpensive strip method. As a result the Donetsk Basin's share of the total Soviet coal output began to fall. By 1940, it had declined to 51 per cent and dropped further because of extensive damage during the 1941–44

German occupation of Ukraine. Yet it still remains the largest field in the Soviet Union, providing about 50 per cent of the USSR's coking coal, a vital ingredient in the production of steel.

Although the Donetsk coalfield has enough reserves to last another 100 years at the current rate of exploitation, it possesses several major disadvantages. First, not only is most of the remaining coal contained in seams less than 1.5 metres thick, but the seams themselves are often steeply inclined. The lack of equipment to resolve this predicament means that a large amount of waste material is mined along with the coal, leading to an increased ash content. In January 1983, a USSR Deputy Minister of the Coal Industry complained that the country was unable to meet its production target for coal because of the declining quality of the coal being extracted with its "constantly increasing ash content."[3] Given a choice between producing machinery capable of mining such seams or concentrating on strip mining in Siberia, Soviet officials are finding the latter alternative increasingly attractive.

Second, the Donetsk mines are very deep. A report of September 1981 stated that 27 per cent of mining in the coalfield was carried out at depths of about one kilometre and that by 1990, most of the seams being worked are expected to be at depths of 1,200 to 1,600 metres.[4] In a speech of March 1986, O. Liashko, the Chairman of the Ukrainian Council of Ministers, stated that the new seams of coal that were being exploited in the coalfield were at depths of 1,000–1,200 metres.[5] Moreover, a report from the former Minister of the Coal Industry of the Ukrainian SSR, Mykola Hrynko, in the summer of 1985 noted that the depth of the Donetsk coalmines increases by 10–15 metres a year, and that one mine in the Torez region is already extracting coal at a depth of 1,319 metres.[6]

Deep mines involve correspondingly greater safety problems, since the excessively hot conditions lead to higher outflows of gas and other potential health hazards. As long ago as 1968, Vladimir Klebanov, who was then a shift foreman in the Donetsk coalfield, refused to send miners to work at the pitface because of their inadequate safety equipment. Mining in the Donetsk region is becoming ever more dangerous and yet the miners are obliged to put in seven-day weeks and long daily hours just in order to maintain current output levels.

There have been a number of debates about the future of the Donetsk coalfield, which ultimately influenced the decision to boost nuclear energy production in the USSR, using Ukraine as the main base of development. The prospects for the Soviet coal industry as a whole were considered reasonable in the mid-1970s, when a forecast of 805 million tonnes was made for the end of the Ninth (1976–80) Five-Year Plan. By 1980, total output of coal was only 716 million tonnes, but once again the prognostications were optimistic at 775 million tonnes.[7] About 725 million tonnes were actually obtained, and the new target for coal output in 1990 is the same as the original 1986 target.

But while the outlook for the coal industry as a whole and for the Donetsk coalfield in particular seems gloomy, Ukrainian officials have long pleaded its case strongly and eloquently. In the winter of 1984, the debate was carried over onto the pages of *Izvestiia* when Coal Minister Hrynko argued a case for the future of the Donetsk coalfield.[8] He declared, for example, that fully 67 per cent of the Donetsk reserves have yet to be developed, a total of 37.5 billion tonnes of fuel. And yet Siberian coal was now said to be "obtruding" into Ukraine's power enterprises because the Donetsk mines were "not being given the opportunity" to raise substantially the output of coking coal.

This lack of opportunity was a direct result, according to Hrynko, of a shortfall in investment: a failure to replenish the mining fund, or to provide new equipment for old mines that have been left for two decades without such attention. The old Donetsk, in his view, had been neglected by central planners, and the implication was that these same planners had followed the advice of specialists in concentrating on the use of Siberian rather than Donetsk coal in Soviet industry.

Having made a case for the Donetsk, Hrynko then embarked on a very realistic critique of the Siberian coalfields, one that has subsequently been echoed in a number of Soviet publications that have tried to ascertain the reasons why with all their natural advantages the Siberian fields have been less productive than envisaged.

First, the Siberian coalfields were lacking in construction personnel—about 110,000–140,000 extra people were required there. Second, the eastern field was lacking a "material-technical base" for the construction workers in addition to a social infrastructure, which was going to cost "billions of rubles." Third, the transportation of coal from fields such as the Kuzbass (Siberia) to the industrial and population zones of the European USSR would necessitate an enormous development of the existing railway system, the costs of which might be in the region of 2.7–3.6 billion rubles. When transport costs are taken into account, Hrynko pointed out, there would be no difference in extracting coal from the Donetsk field, in spite of the obvious mounting geological difficulties in Ukraine. This remark was given further backing by a *Radio Moscow* broadcast (24 January 1986), which observed that 5 million tonnes of coal a year are lost during transport by railway.

Hrynko's pleas were in vain, however. In late October 1985, he was removed from office.[9] In December, long-time Coal Minister of the USSR, Borys Bratchenko, who had often lent strong support to Hrynko, was also dismissed, and sent into retirement.[10] The overhaul of the leadership of the Soviet coal industry reflected the dissatisfaction of the

Soviet leaders, although it did not necessarily spell the end of coal as a leading source of Soviet energy.

But how viable are the Siberian and the other non-European USSR coalfields as alternative suppliers of Soviet energy needs? In 1980, the future of the opencast coal mines of the East was being painted in glowing colours somewhat akin to those used to describe the future of nuclear energy in the USSR today. In September 1980, for example, *TASS* noted that over the past few decades, the fuel balance of the USSR had moved sharply in favour of oil and gas, whereas coal's share had fallen to 27 per cent. Yet the outlook for coal was considered good. Oil and gas were to be used to a greater extent as technological raw materials, and this would lead to the increased importance of coal in the fuel balance, primarily the coal mined from opencast workings (strip-mining). In a statement of 14 November 1980, *TASS* foresaw that coal produced by the opencast method, which at that time accounted for 36 per cent of the total Soviet coal output, would rise to 50 per cent "in the near future."

In June 1982, *Pravda* noted the importance of the Kuznetsk coalfield of Siberia in overall Soviet coal production. It declared that in the year 1982, this coalfield alone would account for almost 150 million tonnes of fuel, including 60 million tonnes of highly valuable coking coal. This Basin, said *Pravda*, is "increasingly" becoming the chief supplier of raw material for the metallurgical industry.[11] As the latter industry is based in Ukraine and traditionally has relied almost exclusively on Donetsk coal, this was a clear indication of a major move from the Donetsk to the Eastern coalfields in Soviet coal production.

This move seemed logical, given the unparalleled geological difficulties facing exploitation of the Donetsk. According to *Pravda*, labour productivity at the opencast mines was three times that of underground mines, while the unit cost of the fuel extracted was 50 per cent less. The newspaper added pointedly that the growing requirements of the country, "especially the European part," for high-quality fuel could only be satisfied by means of the development of the Kuznetsk coal basin. Even in the 1975–80 period, the share of the Donetsk coalfield as a supplier of coal for the central electric power stations of the USSR had begun to decline, from 64.8 million to 59.7 million tonnes. Simultaneously the share of the Siberian and Kazakhstan coalfields rose. The Ekibastuz coalfield of northern Kazakhstan supplanted the Donetsk as the largest supplier of coal to Soviet power stations during this period, with 62.4 million tonnes, up from 44.1 million in 1975.[12]

In September 1985, *TASS* declared bluntly that "the main centre of coal production in the USSR is moving to Siberia." In 1985, it noted, the share of opencast-mined coal in the total Soviet output was 42 per cent, but by the year 2000 it would reach 60 per cent.[13] The "Basic Directions of the 1986–1990 (Twelfth Five-Year) Plan" also envisaged that the plan to raise coal output from its current level of around 725 million tonnes per annum to 780–800 million tonnes by 1990 would be met by accelerating the exploitation of the Kuznetsk, Ekibastuz, Kansk-Achinsk and other coal basins of Siberia and the Far East, and that opencast coal would already account for 48 per cent of Soviet coal output at the end of the Twelfth Five-Year Plan.[14] Thus the Soviet authorities foresaw the eventual dominance of the Eastern coalfields in coal production, and since over 50 per cent of Soviet coal is used to produce electricity, Siberia had an increasing (but not decisive as will be shown below) part to play in Soviet electricity production mainly as a result of the decline of the Donetsk Basin.

In some respects, there was an inevitability about the move away from the European USSR to Siberia in coal production, as in other spheres of natural resources. Siberia and northern Kazakhstan possess over 90 per cent of Soviet coal reserves. The main question was how quickly a significant increase in output could be attained. In 1981, the Soviet authorities felt that production at the Kuznetsk coalfield could be raised from 150 to 250 million tonnes, with an ultimate output level of 550 million tonnes a year.[15] The latter figure would represent about 73 per cent of the current Soviet output. Together with the Kansk-Achinsk coalfield, the Kuzbass makes up the most important Siberian coal region. But despite its enormous coal reserves, the Eastern region of the USSR has posed substantial problems for the Soviet authorities from the outset, and to date has not been a viable alternative to meet Soviet energy demands.

In both Siberia and Kazakhstan, there has been a fundamental failure to establish well-equipped settlements in these remote and climatically adverse regions. In October 1981, for example, the newspaper *Kazakhstanskaia pravda*, in an interview with the First Party Secretary of Ekibastuz City Committee, G.A. Nikiforov, referred to "a great number of blunders and oversights" in the way the coal industry of the region was being established. Construction and repair facilities were said to be in poor condition, the transport system was inadequate, and whereas the city's population had risen by 26,000 between 1978 and 1981, additional housing had been provided for only 13,000 people. The coal industry of the city also required numerous personnel, including 9,500 workers and engineers.[16]

This shortage of personnel was recognized in a Decree of the CC CPSU and USSR Council of Ministers dated October 1981, which examined the future development of strip-mining for power generation as a matter of the highest priority. Starting in 1982, Komsomol and student teams were to be sent to the coal and electricity ministries to work on the opencast mines of the East and to assist in the construction of power stations next to the mines. One of the reasons behind the issue of this decree was the question of transporting Siberian and Kazakhstan coal over great distances to the power stations of the European USSR.[17]

According to the December 1982 issue of *Soviet Geography*, the distances required to transport raw fuel from the East to the European USSR power stations have been increasing constantly. In 1970, the average distance was 861 kilometres; by 1980 it had increased to 923 kilometres. A deficit in the production of steam coal in the Donetsk coalfield had necessitated increased supplies of coal from the Eastern fields, particularly the Kuznetsk Basin. But in 1982, the transport system did not meet the authorities' requirements and there was a shortfall of about 17 per cent in the amount of Siberian coal actually reaching the power stations.[18]

A variety of solutions have been put forward to counter the highly expensive and unreliable transportation of fuel—coal in particular—from the East. One proposes to transform the hard coal into liquid (coal slurry) and to transport it by pipeline rather than railway to the western part of the country.[19] A second, and perhaps more reasonable alternative is to construct power stations in the East that are adjacent or very close to opencast mines, which would cut out the transportation of fuel.[20] But even this option, which has been put into practice in some areas, still entails transporting the electricity across hundreds of miles by means of high-voltage transmission lines to the principal consumers in the western USSR. And even in the European part of the country, the transmission lines seem to be in constant need of repair.

The overriding problem with the Siberian and other Eastern coalfields has been their failure thus far to fulfill their output potential. In an article of June 1982, for example, *Pravda* pointed out that the potential of the Kuznetsk Basin was clearly underexploited, and that for the past five years, the miners had been unable to surpass the 150-million-tonne mark in annual output. Internal reserves, the newspaper stated, were being brought into use too slowly, and equipment was standing idle for long periods because of chaotic organization. The USSR Ministry of the Coal Industry had not made preparations for the opening "of a single new mine" in this region, "even though capital outlays here double or triple that of any other basin." Huge coalmines existed, but had neither the men nor the equipment to operate them. Mining equipment was said to be seriously deficient and in the Kuznetsk coalfield "there is virtually no place where bulldozers, excavators and traction motors for diesel-electric locomotives can be reconditioned."[21]

Has the situation improved since then? Evidence suggests that in 1986, at the start of the Twelfth Five-Year Plan, the Kuzbass mines were still falling well short of their planned targets. In a broadcast of 16 January 1986, *Radio Moscow* verified that "the coal miners of the Kuzbass mining region in Western Siberia have not met plan targets in recent times.'' Later in this same month, officials of the USSR Ministry of the Coal Industry focused on ''existing shortcomings'' in the industry and singled out the Kuzbass coalfield in this respect.[22] Finally on 11 February 1986, *Radio Moscow* stated that ''for a long period, the miners of the Kuzbass have been unable to raise annual output above 150 million tonnes,'' and moreover that over the past year, that figure had actually *decreased*. It laid the blame for this state of affairs on a variety of factors, including water in the mines, transport problems and excessive manual as opposed to machine labour in the mines.

Social factors, however, may also have had a major role to play in the disappointing results attained recently in the Eastern coalfields of the USSR. Most important has been the evident reluctance of Soviet workers to move out to remote areas; and, concomitantly, the failure of the Soviet authorities to provide facilities in distant regions in order to encourage the workers to move there. *Pravda* analyzed the difficulties in one remote area recently: the Buriat Autonomous Republic and Chita oblast, in an article entitled ''Who Will Develop the Resources?''[23] It provided a succinct summary of some of the main difficulties encountered, and went far in illustrating why the ostensibly simple solution to Soviet energy problems—that of developing Siberia—is really not so straightforward at all.

Having referred to the dearth of workers at the mining enterprises in this area, *Pravda* stated that over 60 per cent of the workforce was aged between 35 and 60. As coal miners in the USSR have the option of retiring at the age of fifty, the labour situation was declared to be unsatisfactory. But why were no young people willing to take up a pioneering role in a remote area of the USSR? *Pravda* cited four reasons:

1. The average monthly earnings at Buriat's key mining enterprises had increased more slowly than in other areas of the country.

2. Living conditions were primitive and the area was sparsely populated.

3. There were few residential buildings, including kindergartens, daycare centres, schools, clubs and stores.

4. No vocational or technical schools existed for the coal industry (clearly many young people would need further training), and the only school of note catered to some 30-40 ferrous metallurgy students.

In addition to the above, one could mention the shortage of housing in the eastern mining facilities. In the Kuzbass in early 1986, for example, 300 miners remained on the waiting list for housing, and miners were said to be leaving the region because of the lack of a place to live. The housing facilities had not been built, said *Radio Moscow*, because the construction plan had not been implemented.[24]

Even more serious than the above social problems has been the lack of adequate health-care facilities in eastern regions. In Ekibastuz, for example, there were complaints a few years ago about the lack of hospital beds, half of which were to be found in buildings adapted from other uses. The tuberculosis hospital there had been under repair for the previous eighteen months, and the maternity home for nine months. Because of overcrowding, the inpatient units were hospitalizing only emergency cases, and frequently were unable to take steps to improve the condition of the chronically ill. The city hospital possessed no departments for rheumatology, otolaryngology, urology or endocrinology. According to the newspaper *Meditsinskaia gazeta*, as the polyclinics treated mainly miners and power-station builders, i.e., workers who operated under exceptionally difficult conditions, the *natural result was higher illness and injury rates*.[25]

It is clear from the above examples that both the naturally hostile climatic conditions of the eastern part of the USSR, and the lack of success of the Soviet authorities in overcoming such natural obstacles by preparing properly for the needs of workers in these regions has rendered the expansion of the Siberian and other coalfields to meet current energy needs a somewhat uncertain undertaking. By the mid-1980s, little success had been achieved. As will be demonstrated below, the diverse problems in both the Donetsk and Eastern coalfields, which intensified during the years of the Eleventh Five-Year Plan (1981–85), made it unwise to be overambitious in the newly proposed Twelfth Five-Year Plan. In fact, the Soviets' emphasis is upon *maintaining* rather than raising dramatically the current rates of production. The only ambitious part of the plan concerns the reduction of the size of the workforce required. Greater emphasis will be placed on technology and machinery in the extraction of coal but such processes have also had more than their share of problems.

In an article of 1982, the British weekly, *The Economist*, stated that while capital costs in the development of coal are high, there are enough proven reserves in the world to last another 225 years at the current rates of consumption, which renders the fossil fuel one of the longer-term nonrenewable resources. In contrast, the future of oil was perceived to be only about 35 years and that of natural gas 50 years.[26]

In the USSR, coal accounts for somewhat less than 25 per cent of the USSR's energy fuels demand, but again its future seems assured in the long run, and Soviet leaders have constantly reiterated their faith in coal, either as the main substitute for natural gas in thermal power stations, or as a resource that would eventually be returned to its rightful place as the primary energy fuel.

There is little doubt that at the present time, both in the USSR and the world, coal has taken a backseat to oil. In 1980, according to the International Labour Organization, oil supplied over 50 per cent of the energy needs of the main industrial nations. At the same time, the ILO acknowledged that this share would fall in the years ahead and that by the year 2000, coal would again be the leading fuel by a margin of 37 to 33 per cent. Given another three decades, the prediction went, coal would account for 33–38 per cent of world energy needs, oil would fall to about 18 per cent, but *nuclear energy would make up 26–28 per cent of the total*. This is a fairly accurate reflection of the thinking of the Soviet leaders also, except that the rate of expansion of the nuclear field is expected to be considerably faster than the above 45-year period of development.

There are two reasons for such rapid growth in the nuclear sphere, in addition to the decline of the European coal industry and difficulties in coal mining in Siberia as mentioned above. Both are concerned with the oil industry. The first is that the Soviet authorities are seeking to preserve their oil reserves as an important future source of hard currency. Rather than export oil to East European countries, for example, the USSR is beginning to investigate the alternative policy of conserving supplies of oil and exporting electricity instead. Second, however, the Soviet oil industry's annual growth rate has slowed down alarmingly. Since Gorbachev became CPSU General Secretary in March 1985, and especially in the first several months of 1986, the Soviet authorities have been preoccupied with problems in their oil sector. Before the Chernobyl accident, the oil industry was the prime economic concern of the Soviet leaders, at least according to the Soviet press and journals.

Before discussing briefly the recent problems in the Soviet oil industry, let us place matters in perspective by looking at the annual rates of increase in output of oil (including gas condensate). In 1960, total oil output in the USSR stood at 147.9 million tonnes. By 1970, it had risen dramatically to 353 million tonnes, which represented an annual average increase of 23.8 per cent. In 1975, the total was 490.8 million tonnes, i.e., the annual average increase in output over that five-year period was an even higher 27.8 per cent. At the end of the Ninth Five-Year Plan in 1980, another spectacular growth in output brought the annual total output to 603.2 million tonnes (24.6 per cent per annum growth). Thereafter, the industry experienced a stagnation in growth. Between 1983 and 1984, oil output decreased for the first time in Soviet history, from 616.3 to 612.7 million tonnes, and the 1984 figure represented a percentage aggregate increase of only 1.5 per cent over a four-year period—0.37 per cent per annum as opposed to the highest annual average increase of 27.8 per cent.[27] In Soviet terms, this represented a major setback and Gorbachev has made it his business to ascertain the reasons for this failure to maintain growth rates (even though the total output at 612–616 million tonnes per annum is the highest in the world).

In September 1985, Gorbachev made a much publicized visit to the main Tiumen oilfield in Western Siberia. There a long list of problems was outlined: violations of economic plans; wasteage; low-quality work; a lack of attention to the enviroment; a shortage of housing for oilworkers; low quality and unreliable equipment.[28] Gorbachev himself made a major speech on 6 September.[29] Having outlined the importance of the Tiumen field in the USSR's Energy Programme, he proceeded to elaborate on the shortcomings there:

> The CPSU Central Committee is worried by the fact that for the third year the Tiumen area is not fulfilling plans for the extraction of oil....The problems accumulated gradually over the years. Today we must say firmly to ourselves that the extraction methods that were envisaged during the first stage of the formation of the oil extraction complex on the Ob have in practice exhausted their potential....The epoch...of "easy oil"...is coming to an end.

Gorbachev laid the blame for the reduction in increase of output on geologists, who had failed to make provisions for the future as a result of complacency; machine-builders, who "have not responded properly to the problems of our oil and gas workers"; and the lags in the development of the power industry. There remained in his view substantial problems both in capital construction and in housing for the construction workers themselves:

> The state of affairs in capital construction is holding back the resolution of many important questions, and I would say is evoking certain concern....The matter, first and foremost, is that there is not enough highly productive machinery among the construction workers. The second thing...is that...the construction workers are in the worst position concerning housing....In the plans for the Twelfth Five-Year Plan, the volume of housing construction is less than in the Eleventh Five-Year Plan.... The result is illogical and out of step with the scale of the work.

The extraction process was becoming more difficult as the more accessible oil reserves were used up, and yet the workers did not even have the basic accommodation to keep pace with the new demands that were being placed upon them. Geologists were not exploring new fields to a sufficient degree and a number of related industries were failing to provide the required assistance to the oil industry.

47

After Gorbachev's visit to Tiumen, the situation hardly improved. In October 1985, the 73-year-old USSR Minister of Petroleum Mining and Petrochemical Industry, Viktor Fiodorov, was replaced.[30] And at the end of this month, the Tiumen oblast party committee held a plenum, which again referred to the deplorable housing situation in the oilfield and the fact that "the style of work of some managers still does not come up to the party's modern demands."[31]

On 24 December 1985, *Pravda* cited another oblast committee plenum in Tiumen, which indicated that the situation in the giant oilfield was very serious.

> There are still a considerable number of [shortcomings]. Thus, planned growth rates in labour productivity are not being sustained in the oblast, return on investment is falling, a serious lag in oil extraction estimated at tens of millions of tons is permitted, and many social issues are being resolved too slowly.

Over the 1983–85 period, declared *Pravda*, "several hundred" leading party personnel in the oilfield had been replaced as being poorly educated and too immature for the positions of trust they held.

In mid-February 1986, the oil industry of the USSR was again made the subject of major attacks in Soviet newspapers. *Sotsialisticheskaia industriia* laid the blame for yet another shortfall in oil production—for the month of January 1986—on the inefficient shipping of equipment to the West Siberian oilfield. It described what it called a "catastrophic situation" because of a shortage of cranes, pipe carriers and other unloading machines at the vast oil complex.[32] *Pravda* stated that 13 per cent of the oil wells in the Tiumen region were idle at the beginning of 1986.[33]

Perhaps the clearest account of the entire situation in the Soviet oilfields was that provided by V.A. Dinkov, the USSR Minister of the Oil Industry at the Twenty-Seventh Congress of the CC CPSU on 1 March 1986.[34] His report went as follows:

> The opening up and utilization of major deposits in the Urals and Volga region and Western Siberia, and the use of new technology enabled the Soviet Union to take first place in the world [in oil production] more than ten years ago and to remain firmly in that position....Comrades, when we assess soberly the state of affairs in our industry, we see our deficiencies, unexploited potential and unsolved problems. In the past decade, the oil industry has developed rapidly; but oil machine-building, capital construction, extraction machinery and technology have lagged behind seriously in their development. Extraction growth rates have been well ahead of geological prospecting workrates. There has been a failure to appraise and take

into account in good time the fact that while industrial oil reserves have been increasing constantly, their quality has deteriorated. This, in turn, has led to an increase in proportional expenditure of all types of resources.

The age-old problem of housing facilities for workers remained, however. Dinkov acknowledged the fact without providing any way to overcome the dilemma.

We have before us the top priority task of constructing at a rapid rate housing, kindergartens, schools and hospitals; entertainment and sports facilities, and buildings for retail trade and agriculture. Above all, we need to do this in Western Siberia...we must improve the living conditions of the oil workers without which the opening up of new deposits is inconceivable.

Why had this not been achieved in the past? Dinkov's reply was that individual leaders bore the main responsibility:

Conditions of oil industry development have changed but the way of thinking and attitudes on the part of the Ministry and many enterprise leaders have remained inert and dominated by past successes. Unfortunately both the Gosplan [State Planning Committee] and State Committee for Science and Technology did not react in time to the changes that were taking place. All these negative phenomena led to nonfulfillment of the Eleventh Five-Year Plan period in Tiumen oblast and in the industry as a whole.

These problems in the oilfields of Siberia—the field that largely accounted for the spectacular growth rates in the oil industry in the 1960s and 1970s—are longstanding now, and are not likely to be resolved overnight. One cannot logically speak of a crisis, given the total volume of production. But the deficit in terms of the plans is significant, about 12.2 million barrels per day as compared to a planned level of over 12.5 million barrels. Further, the costs of producing oil in Siberia are much higher than other main oilfields in the world. For example, they are up to six times as high as those in the Middle East.

In addition to the costs of production, falling world oil prices may have affected the relationship between the Soviet Union and its East European neighbours, which have relied heavily on inexpensive imports of Soviet oil for their energy needs. It is plausible that if world oil prices continue to fall in the immediate future, then the East European countries will be paying more than the world price for Soviet oil. This, in turn, might cause these states to consider importing cheaper oil from the Middle East.

In contrast to oil, Soviet gas production is enjoying a boom period, and it seems likely that the Soviet authorities will try wherever possible to conserve oil supplies, replacing oil with coal and gas for domestic consumption. As for East European states, it is probable that the USSR will in the long-term cut back on oil exports in line with the policy espoused by the Twenty-Seventh Congress of the CC CPSU to *conserve* oil resources in the USSR.

Thus for a variety of reasons that are linked to very specific difficulties in the oil industry, the Soviet authorities prognosticated only a very modest increase in oil production for the 1986–90 period of the Twelfth Five-Year Plan up to 630–640 million tonnes.[35] There are major problems to be overcome and new geological surveys to be undertaken in the remote areas of the northern and far eastern USSR. The climate, the lack of facilities for workers, the costs of extraction from new, less accessible fields, and the perennial weaknesses of the Soviet railway transport system mean that for some years, the stagnation in the Soviet oil industry is likely to continue.

The hazards of obtaining energy resources from Siberia have a direct effect on Soviet Ukraine, which is a major consumer of energy. At the same time, Ukraine's geographical location on the western borderland of the USSR makes it a suitable area from which to forge closer economic links between the USSR and its East European neighbours. Even in the 1970s, the Soviet authorities were seeking alternative supplies of energy, and moreover supplies that could be more or less guaranteed: that did not depend on a Donbass miner working every weekend of a month, or on the Soviet railway or supply system to a distant oilfield. The demand for electricity was considerable. A means to this end was sought in nuclear energy, which appeared to the Soviet authorities to be an industry that was both economical and reliable.

Nuclear energy was an avenue that had been explored by the Soviet authorities with the establishment of what the Soviets claim was the world's first nuclear power plant for civilian uses at Obninsk in 1954. But it was an option that had not been explored as a *major* supplier of energy. Not only was nuclear energy seen as a way to plug an energy gap, but the exporting of nuclear energy was perceived as the most convenient way to circumnavigate the need for Soviet oil in Eastern Europe.

CHAPTER THREE

Nuclear Energy Development in Eastern Europe

The Chernobyl disaster had a significant impact on the nuclear programmes of East European countries. Substantial problems were in evidence at these plants before the accident. They had intensified in the 1980s as a result of a major programme of expansion that was quite unrealistic if measured by past and current construction rates. Chernobyl also fuelled anti-nuclear movements in some of these countries that have made the long-term nuclear energy plans even more difficult to fulfill.

The USSR, through the Council for Mutual Economic Assistance (CMEA), is playing the leading role in the development of nuclear energy in Eastern Europe. Through the so-called MIR system, there have been attempts to establish a unified grid for all the countries involved, and to develop water-pressurized reactors that are manufactured either in the USSR or, under Soviet supervision, in Czechoslovakia.

Within the USSR, Soviet Ukraine has been assigned the key role in East European nuclear development. Two 750-kilovolt transmission lines connect Ukraine to the East European grid, including one that links up with the Chernobyl station. In February 1985, Radio Prague confirmed Ukraine's role when it referred to the development of an East European nuclear power system that also encompasses Cuba, Yugoslavia and "the western part of Ukraine." Similarly on 22 August 1985, officials from Bulgaria, Poland, Czechoslovakia, Hungary, East Germany and the USSR chose the West Ukrainian city of Lviv as the location for discussions on energy supply questions. The meeting was timed to coincide with the opening of the transmission line linking Ukraine and southeastern Poland for the export of electricity from the USSR, but the location was of significance.

The development of nuclear energy in the USSR, and especially in

Ukraine, cannot be dealt with adequately without looking at its progress in Eastern Europe. In a variety of ways, Soviet Ukraine has been linked intricately with East European countries as a result of its geographical location on the western borderland of the USSR. Of the four Ukrainian nuclear plants operating in 1985, two were exporting electricity to Romania, Bulgaria, Poland and Hungary, while a third (Khmelnytsky) also for East European needs was scheduled to come on-line in 1986.

East European involvement in Ukrainian nuclear construction will be dealt with in the next section. Suffice it to say here that Poles, Romanians and other nationalities are currently working on the construction of Ukrainian nuclear plants, while Ukrainian personnel — usually skilled engineers — are involved in the design and operation of some East European nuclear stations.

In theory, the CMEA system works on a co-operative basis, with each country contributing different resources to the overall operation. The premise is that with the exception of the USSR itself, the East European countries lack the technological resources to develop nuclear energy individually. At the same time, it is considered by some governments as the only means of averting a future energy crisis. Bulgaria, Romania and Poland in particular are not well endowed with natural resources. Thus Romania now contributes water tanks for CMEA nuclear plants; East Germany cranes; Hungary refuelling and water treatment equipment; Bulgaria pumps and condensers; Poland pressurizers and heat exchangers; Yugoslavia cranes, pumps and some of the components for graphite-moderated reactors (in the USSR); while Czechoslovakia, which is playing the second most important role after the USSR, manufactures water-pressurized reactors at its Skoda factory, in addition to steam turbines, generators and other equipment.[1]

It should also be mentioned at the outset that nuclear energy development has given the USSR more control over and increased input into the economic programmes of its neighbours. With the exception of that being constructed in Romania and the one in operation in Yugoslavia, all the reactors being used in Eastern Europe are of Soviet or Czechoslovak make. All the CMEA countries also rely largely on Soviet technical knowhow and, quite often, Soviet personnel on the spot. Of all the countries involved in the CMEA nuclear programme, only Czechoslovakia has the developed infrastructure to be able to operate individually, and even the Czechs still rely heavily on auxiliary equipment from other countries involved. The whole programme has thus given the USSR a great deal of leverage over its allies, and this leverage is increasing as the programmes develop.

An illustration of Soviet hegemony over East European nuclear energy development is provided by the recent activities of A. Antonov, in his position as Chairman of the Intergovernment Commission for the Atomic Equipment of CMEA countries. In April 1985, Antonov was in Czechoslovakia, where he signed a Protocol with the Czechoslovak representative that outlined the development of nuclear energy for the 1986–90 period in Czechoslovakia.[2] In the following month, Antonov was in Sofia, where he met Bulgarian leader Todor Zhivkov, to discuss Bulgarian-Soviet co-operation in the construction of the Bulgarian nuclear plants at Kozloduy and Devnia.[3] Although CMEA sessions have been held in all the countries involved, it is clear that the Soviet representative is the principal, if not omnipotent, authority. On 26 June 1986, *Pravda* published Mikhail Gorbachev's speech, delivered in Ukraine, which called for the "deepening economic co-operation and economic integration" of the USSR and its trading partners, a remark perceived by some Western reporters as a demand for greater co-ordination in nuclear energy with more emphasis on Soviet priorities.

Co-operation between the CMEA countries in the development of nuclear power first began in 1960, although a comprehensive programme was only developed in 1971, which anticipated economic integration for the next 15–20 years, including the nuclear sphere. In 1972, a Standing Commission for Nuclear Energy was created, which divided up research responsibilities, with most of the chief parts for the industry being manufactured by the USSR. At this time, the *Interatominstrument* organization was created. In the following year, the *Interatomenergo* organization was founded for the purpose of integrating nuclear energy plans by using standardized reactors in the CMEA countries and Yugoslavia. Subsequently, short and long-term plans were created, and the construction of water-pressurized reactors took place at an ever-increasing rate.

Generally, progress has been slower than anticipated, both in the Soviet Union and Eastern Europe. In 1985, in the CMEA countries, nuclear plants accounted for only about 8 per cent of the electricity generated, mainly as a result of delays in putting nuclear plants into operation.

In October 1985, the Forty-Ninth Session of the CMEA Standing Committee on Co-operation for the Peaceful Use of Atomic Energy was held in Moscow, attended by delegates from the USSR, Bulgaria, Hungary, Vietnam, East Germany, Cuba, Poland, Romania, Czechoslovakia and Yugoslavia. According to a Secretary of the CMEA, V. Sychev, the total capacity of nuclear plants in the CMEA was to be doubled by 1990, and the Session adopted a comprehensive programme to the year 2000. On 22 December 1985, Moscow television interviewed USSR Minister of Power and Electrification, A.I. Maiorets, who confirmed that:

Nuclear power plants are being built with our participation, with our technical guidance and according to our designs in virtually all the CMEA countries: in Bulgaria, Hungary, the CSSR [Czechoslovakia], the GDR and Poland, and a decision has just been taken to begin preparations for the construction of a nuclear station in Yugoslavia.... This big programme is realistic because all CMEA countries have pooled efforts to produce unique power engineering equipment.

At the end of 1985, according to the report of the International Atomic Energy Agency (IAEA—issued in April 1986), the share of nuclear power in the electricity generated by the CMEA countries and Yugoslavia, in percentages, was as follows: Bulgaria 31.6; Czechoslovakia 14.6; East Germany 12.0; Hungary 23.6; Yugoslavia 5.1 and the USSR 10.3. Neither Poland nor Romania had nuclear plants in operation, although construction has been under way for years in both countries. Clearly the above percentages have not satisfied the Soviet authorities since they still entail major dependence on imports of Soviet electricity on the part of most of the countries involved. Thus a crash programme has been initiated that is unprecedented in its scope and speed. Since development in each CMEA country has been so varied, a brief review of the development of nuclear energy in individual countries will follow.

Bulgaria has four 440-megawatt water-pressurized reactors in operation at its Kozloduy plant, the first of which began generating in July 1974, with the second following in November 1975, a third in January 1981 and a fourth in June 1982. Work on the fifth reactor is in progress. This is to be a larger 1,000-megawatt reactor, which is being designed with the aid of specialists from the Zaporizhzhia nuclear power plant in Ukraine. Computer equipment is being installed by a joint Bulgaro-Soviet group of specialists. A sixth reactor is said to be under construction in the vicinity.[4] In the construction process, guest workers are being used from a variety of countries, including Poland, Vietnam, Cuba and Ethiopia.[5]

In addition to the Kozloduy plant, which is expected to be fully operational by 1989, the Bulgarians are constructing another nuclear plant near the town of Belene on the Danube, where two 1,000-megawatt reactors are expected to be generating by 1991. In short, then, capacity at Bulgaria's plants is to be more than tripled over the next five years, from 1,760 to 5,760 megawatts. The reasons are twofold: first, Bulgaria lacks natural resources to service its power industry; second, the country consumes an amount of electricity per head of population that is excessive.

According to the Bulgarian Minister of Power Engineering, N. Todoriev, consumption of electricity per capital of population in Bulgaria is more than 2.5 times the world average. Todoriev stated that Bulgaria is still "lagging behind in the rational use of fuel and energy, and in raising the energy efficiency of the economy."[6] The authorities announced in May 1986 that Bulgaria would not generate enough electricity in 1986 to meet the country's needs; that an extra 400 megawatts would be required for new industrial facilities, and 250 megawatts to provide electricity for 50,000 new apartments. Consequently strict rationing of electricity in the country was renewed.[7]

The solution to this problem, according to the Bulgarian and Soviet authorities, is first, to bring the fifth Kozloduy power block into operation by the end of 1986, and second, to raise the share of nuclear energy in electricity production to 60 per cent by the year 2000, which would make Bulgaria one of the world's heaviest users of nuclear energy per head of population. Again, the Soviet Ukrainian connection should be emphasized. Ukrainian engineers are working on the construction of the two Bulgarian stations, which are being built on the Zaporizhzhian model.

Of all the East European countries involved in the CMEA nuclear energy programme, Czechoslovakia appears to be the most committed and the most developed outside the USSR. Somewhat ironically in the wake of Chernobyl, one of the main reasons for the extensive nuclear development has been protection of the environment. Czechoslovakia's power industry has long been dependent upon supplies of lignite coal, which give off sulphuric acid fumes that constitute an environmental hazard. In contrast, nuclear plants are considered virtually pollution free.

The Czechoslovak programme is taking place under the close and constant scrutiny of the USSR, which appears to have been involved at almost every stage of development. Most recently, the Eighth Session of the Czechoslovak-Soviet plenipotentiaries for developing nuclear power in Czechoslovakia up to the year 1990 ended in Moscow on 10 April 1985, with the signing of a Protocol between Aleksei Antonov and the Czechoslovak Deputy Federal Premier, Ladislav Gerle.[8]

Czechoslovakia's first nuclear power plant was constructed near Jaslovske Bohunice in western Slovakia from 1974 onward. The first 440-megawatt unit began generating in 1978, with a second coming on-line in 1980 and a third in 1984. With the completion of the fourth unit in 1986, it is estimated that this plant alone will save the country 10 million metric tonnes of brown coal per year. In August 1985, a second plant came into operation at Dukovany in southern Moravia, and the second energy block was started up here early in 1986, evidently ahead of the official schedule, which was timed to coincide with the Czechoslovak Party Congress in March 1986.[9] A third unit was also scheduled to come on-line later in 1986.[10]

Altogether in Czechoslovakia, six reactors are in operation at two plants while a further ten are under construction at Dukovany and at two new locations in Mochovce in western Slovakia, based on VVER 440 reactors, and at Temelin, which is only about sixty kilometres from the Austrian border, using VVER 1000 reactors for the first time in domestic industry. By 1990, if plans are met, ten reactors should be generating electricity, at which time the share of nuclear power in electricity output will have risen from 20 to 28 per cent. By the turn of the century, it is anticipated that well over 50 per cent of Czechoslovakia's electricity needs will be met by nuclear energy, and that the figure will be around 73 per cent by the year 2020. Thus like Bulgaria, the nuclear option is regarded as the preferred path for future energy development.

Czechoslovakia is also a major participant in the manufacture of nuclear equipment, both for domestic use and for export. According to the *Journal of Commerce* (1 May 1986), the Czechoslovak energy programme will account for 37 per cent of all industrial investment in the 1986–90 period. The Skoda works produces water-pressurized reactors of both 440 and 1,000-megawatt variety, and is the only CMEA reactor producer outside the USSR. Yet the Czechoslovak nuclear industry has been beset by major problems, which make the rapid expansion outlined for the future quite improbable in terms of adhering to the proposed plan, quite apart from the repercussions of the Chernobyl accident.

In August 1985, for example, the Czechoslovak Fuel and Power Minister, Vlastimil Ehrenberger, announced that the country's nuclear power plants would not be able to fulfill their targets for 1985. Plant construction, he said, was being delayed by numerous shortcomings, including the supply of materials and spare parts.[11] It is believed that the completion of construction work at the Jaslovske Bohunice and Dukovany plants has been delayed for over a year because of these supply defects.[12] In October 1985, it was again reported that shortages of goods and installations were threatening the construction of the Dukovany plant and that winter preparations for the third and fourth blocks were "a complex matter." Problems were being encountered with construction and assembly operations, and with the installation of the heating system at the plant. Some of the equipment required evidently had not reached the Dukovany site at the required time, notably pumps from the Sigma Company of Lutin, and valves and vents from a firm in Usti-nad-Labem.[13]

Maintenance at nuclear plants was the subject of criticism by Czechoslovak energy officials in early spring 1986, at which time the overhauls of two operating reactors were recommended, one at the Bohunice plant, the other at Dukovany.[14] Given the evident seriousness of these problems, and the frequency with which they are highlighted in Czechoslovak news services, one must have doubts about the future programme that foresees an increase in nuclear capacity from 2,200 to almost 5,000 megawatts by 1990, and to over 10,000 by the turn of the century.

Of the East European countries, Hungary appears to be the most directly linked to Ukraine in nuclear energy development. The Paks plant, which is Hungary's only nuclear plant to date, is being designed by the Kiev branch of the recently established Institute for the Planning of Thermal Nuclear Energy (*Atomteploelektroproekt*), which specializes in using nuclear energy to supply heat and water to major towns. The Kiev engineers reportedly are supervising the construction process and resolving any technical questions that arise.[15]

The Paks plant, which is located in southern Hungary, first began to generate electricity in the autumn of 1983. A second 440-megawatt reactor came on-line in 1984, and two more are planned for the early years of the 1986–90 plan. At full capacity (1,760 megawatts), the plant should make up about one-quarter of Hungary's total electricity supply, a saving annually of 2.5 million tonnes of oil, or 8 million metric tonnes of coal equivalent. Hungary has long had problems in producing sufficient quantities of coal, and the Paks plant was actually commissioned as long ago as 1965. According to the Hungarian authorities, the country possesses enough uranium to supply the plant until the year 2020. The uranium ore, which is to be found among coal deposits near Pecs, is said to be sufficient to fuel a second nuclear plant in Hungary, the construction of which is to begin before 1990.[16]

According to a Western authority on Hungarian affairs, the Paks plant has been plagued by last-minute design changes, cost overruns, waste and a dissatisfied workforce. He notes that although at a February 1986 meeting, Hungarian and Soviet officials decided to double the capacity of the plant from the size originally anticipated, the costs of building Paks have now tripled, and official statements about savings as a result of economy of fuel are no longer applicable. This same source states that there have been labour problems at the plant, resulting partly from a shortage of workers in the early phases of construction, and partly because of a decision to abandon the use of two separate contractors (the Power Plant Investment Enterprise and the Paks Nuclear Power Plant Enterprise) in favour of total control by the latter. Evidently this move led to bad feeling, bickering and disorganization of work at the Paks plant.[17]

Despite the labour setbacks at Paks, the plant will continue to expand to huge proportions. Six reactors are now envisaged there with an ultimate capacity of 5,760 megawatts. With the construction of a second nuclear plant in the 1990s, the share of nuclear power in Hungarian electricity production will rise to 40 per cent by the end of the century if targets are met.

Information about Romania's nuclear energy plans is not easy to obtain. Initially, it appears that the country's first nuclear plant was to be established in Olt county, an area which one Western analyst has described as totally unsuitable but for the fact that it is the native county of President Ceausescu,[18] and subsequently abandoned, only to be taken up a second time quite recently. Also, despite close co-operation with the USSR in making its nuclear plans, and in participating in the Soviet construction at the South Ukrainian nuclear plant in Mykolaiv oblast, Romania opted initially to use Western technology in its nuclear industry. Consequently, the first plant, which is still under construction at Cernavoda on the Danube, began in 1981 using Canadian technology and CANDU reactors. The first of three reactors was scheduled to come online in 1985, but the Romanians have fallen behind schedule.

Romania has suffered from a shortage of electricity for a number of years. The agreement with Canada represented a major effort to overcome this deficit. On the basis of a 1977 meeting, according to a Romanian source, Canada was to deliver equipment for the plant's operation and to train Romanian specialists in relevant Canadian units. In 1980, another agreement was drawn up, allowing for "technical exchanges" between Romania's State Committee for Nuclear Power and Atomic Energy of Canada Limited (AECL). The total capacity of the Cernavoda plant was to be 3,500 megawatts, based on reactors of 630–650 megawatts capacity.[19]

In February 1986, the Romanian authorities confirmed that for the construction of its second plant, it will rely on CMEA technology and input. Work on a new plant, near the city of Piatra Neamt in Moldavia, began in March 1986. It is reported that Czechoslovakia is preparing three reactors, each of 1,000-megawatt capacity, for the plant at very favourable terms for the Romanians. The design is in the hands of the USSR, which is also supplying the bulk of the equipment and the skilled personnel. The first power block is scheduled to begin generating in 1991.[20]

According to a report from the Yugoslav news agency, Romania produced 70 billion kilowatt/hours of electricity in 1985, which fell far short of the country's requirements. More significant, imports of electricity from the USSR and Albania were evidently insufficient to make up the deficit.[21] Nuclear power is badly needed in Romania, but thus far the nuclear programme appears to be in some disarray.

The failures in Romania resemble those in Poland, where the Zarnowiec nuclear plant in Gdansk province was originally scheduled to start operating in 1984, according to a PAP (Polish news agency) statement of early 1980. Later this date was put back to 1985, and then in June 1984, PAP announced that efforts were being made to ensure the generation of electricity at Zarnowiec in 1991, i.e., seven years behind the original schedule.[22] No reasons were given for this long delay, but evidence indicates a strong current of domestic opposition to the construction of nuclear plants in Poland. Early in 1986, for example, the *Zycie Warszawy* newspaper included a discussion about the Zarnowiec plant, which noted that the plan for the first reactor was "taken" from the more experienced Soviet neighbour. It then focused on local concerns about the plant, at which construction only began after an entire village was moved from Kortoszyno to Odargowo.

The local population, it stated, "did not want to sit on an atomic bomb," and it took officials "some time" to persuade residents that the plant would be as safe as possible. Radioactive water, for example, was to be sealed hermetically, and the radioactivity removed through steaming. Radioactive waste, on the other hand, would be sent to the USSR on the basis of an agreement between the two countries. Control over the environment was said to be "rigorous," and a radiation monitoring station had been established even before building work on the power plant began.[23] Clearly public concern had something to do with the long delay in completion of the plant. But there is also an acute labour shortage at the site. The project appears to be labour intensive and a maximum labour force of 12,000 persons has been projected.[24]

The main Polish input into the plant, aside from the majority of the workforce, is the production of turbines and generators. The reactors for Zarnowiec are to be supplied by the Skoda firm of Czechoslovakia, while the USSR is assisting with (if not supervising) construction work and providing the other nuclear components.[25] Poland's proposed plan for future nuclear development was worked out at a May 1984 meeting between General Jaruzelski and the late CPSU General Secretary, Konstantin Chernenko.[26]

The plan for Zarnowiec is as follows: the first two power blocks are to be working by 1991; with a third coming on-line in 1993, and a fourth in 1994. Total capacity is to be 1,860 megawatts, and two other plants, both with an ultimate capacity of 4,000 megawatts, are to be constructed, one of which will be near Klempicz, northwest of Poznan.[27] While these plans are hardly less ambitious than those of Bulgaria and Czechoslovakia, problems have been intensified in Poland by environmentalist protests and a fundamental opposition to nuclear power development.

In 1981, for example, a site for a new four-reactor plant was selected in the picturesque Karolewo region, near Plock, an area of pine forest and glacial lakes. A strong campaign by local residents against the location received support from Communist officials in the area and from the official League of Nature Protection environmentalist group.[28] A final decision on whether to use the Karolewo site has yet to be made, but for the moment the plans appear to have been shelved.

Other pre-Chernobyl protests were far from uncommon. In April 1986, for example, the Polish news service reported a protest of residents in the Zelenia Goria province close to the East German border against plans to locate a nuclear waste dump in the vicinity. Underground bunkers constructed by the Germans during the Second World War were felt to be suitable repositories for nuclear waste. The protests, which began at in 1985, if not earlier, are being led by the so-called Pron movement, a semi-official body established after martial law with a mission to find common ground between Polish citizens and the Polish government.[29]

In the long term, it appears that the Polish authorities are determined to become more reliant on nuclear energy. The immediate goals are to bring Zarnowiec into operation and to raise the share of nuclear power in electricity production from zero to 16 per cent by the year 2000.

East Germany also appears to be committed firmly to nuclear power, despite a slight reduction in its proportion of electricity generation between 1980 and 1985, from 12 to 10.5 per cent. According to the IAEA materials, five reactors were in operation at the end of 1985, four of the 440-megawatt variety at the "Bruno Leuschner" plant near Griefswald and a smaller reactor put into operation in 1966 at Rheinsberg. While East Germany is a leading lignite producer and can possibly afford to develop more slowly than its CMEA partners in the nuclear sphere,[30] the indications are that at least six new reactors are under construction there, including two 1,000 megawatt reactors at a new location in Stendal, and four 440-megawatt reactors that are being supplied by Czechoslovakia to the Nord plant, all of which should come on-line in the latter part of the 1980s.

Yugoslavia is perhaps the most interesting case-study of the East European countries because of its lack of commitment to, but participation in CMEA meetings and plans. In fact, Yugoslavia's only extant nuclear plant to date was built by the U.S. Westinghouse Company at Krsko, fifty kilometres north of Zagreb in 1983, with a capacity of 632 megawatts.

In July 1985, however, Tanjug announced that the country's electric power industry had initiated action for the construction of four nuclear power plants by the year 2000, with the first to be built at Prevlaka in Croatia, 38 kilometres south of Zagreb.[31] Bids from international firms to build the plant were invited at this time, and by February 1986, twelve companies from France, Italy, Japan, the United States, Argentina, Britain, the USSR, West Germany and Canada had put forward offers.[32]

The international nature of the proposed construction and the rate of expansion of nuclear energy in Yugoslavia generally sparked an intensive debate in scientific and government circles that may have also had repercussions in the CMEA countries, including the USSR. Simply put, the questions raised by Yugoslav scientists before the Chernobyl accident are applicable to virtually all the CMEA countries involved in the ambitious programme for nuclear energy expansion in Eastern Europe and the Soviet Union over the next fifteen years.

The debate reached the higher echelons of the administration with the publication in the February edition of the Yugoslav journal *Komunist* of an article by Professor Slavko Kulic, a scientific advisor at the Zagreb Economic Institute and head of the Centre for Strategic Research. Kulic pointed out what he felt would be the serious consequences for the present and future if the Yugoslav government were to pursue the nuclear route in energy development. According to Kulic:

> The way in which nuclear energy has thus far been produced and used calls for an explanation of the extreme situation that we have been pushed into by people of inadequate knowledge, overexcited technocrats who have chosen to secure their power in society by promoting the nuclear alternative of generating electricity without any consideration of what it would mean for future generations and for us who are using that form of energy....I find it strange that decisions about this issue are being made by people of inadequate knowledge who are committing intellectual violence on the public. The fact that our general public is inadequately informed is...unacceptable from the viewpoint of socialist democracy and the system of socialist self-management.[33]

This strong statement was soon followed by others. In particular, the Yugoslav press was heavily critical of nuclear energy development in the months leading up to the Chernobyl disaster. On 23 February 1986, the Belgrade daily *Politika* carried a statement by a member of the Central Committee of the governing Communist League of Yugoslavia (LCY), Dragisa Ivanovic, asserting with regard to the proposed Prevlaka construction "Let us not allow a noose to be put around Yugoslavia's neck." On the same day, the *Politika Ekspres* newspaper revealed that over 130 mothers had sent a petition to the President of the Assembly of Yugoslavia, signing their names under the slogan "We do not want another Hiroshima in Yugoslavia."

In response to these and the many other high-level protests against Yugoslavia's proposed future nuclear power development, the government of Mrs. Milka Planinc acted cautiously. In March 1986, she stated that Yugoslavia, with its modest oil and gas reserves, could not hope to resolve its energy problems without recourse to nuclear power.[34] At the same time she promised that a democratic debate would be held on the issue, a step considerably more radical than had been taken thus far by any of the countries participating in CMEA schemes. The situation was, however, intensified by the accident in Ukraine.

Impact of the Chernobyl Disaster in Eastern Europe

That the Chernobyl accident will have a major impact on the East European nuclear power programme is clear. According to one Polish-area specialist:

> Few can doubt that the Chernobyl accident will have political consequences in Eastern Europe. At the very least, one may expect an increase in public concern over the planned expansion of nuclear power plants.[35]

Similarly, the Washington-based *Plan-Econ Report* of May 1986, in assessing the consequences of Chernobyl, also declared that the "Chernobyl accident will cause East European countries to pause and reevaluate some of their ambitious nuclear power construction plans, especially in Bulgaria and Czecholovakia."[36] The authors also maintained that reductions of Soviet electric power deliveries to Hungary, Bulgaria and Czechoslovakia could take place.

While it would be misguided to make too much of the protests that have followed the accident, which are essentially small-scale, it is nonetheless clear that the nuclear tragedy had an almost immediate impact, particularly in Poland, which was most directly affected by the radiation cloud, Czechoslovakia and Yugoslavia, where there was already a powerful anti-nuclear lobby in place. It is also clear that these protests have already affected psychologically and otherwise, the major nuclear programme set forth for Eastern Europe. An issue that could determine election results in Western Europe has also swept across Eastern Europe. France, Belgium, Bulgaria and Czechoslovakia have similar commitments to nuclear power. Inevitably, the *form* of the protest against nuclear development has differed, but it is no less significant for occurring in countries with more authoritarian forms of government.

Czechoslovakia was less than forthcoming in its reporting of the Chernobyl accident and, by and large, simply repeated statements made earlier by Soviet news agencies. According to one analysis:

> Czechoslovak behaviour throughout the Chernobyl crisis so far has revealed a certain measure of insecurity glossed over by the veneer of a reassertion of faith in the infallibility of Soviet and East European science and technology. There are two possible explanations for this attitude. First,

Czechoslovak heavy industry has become a major conduit for the diffusion of Soviet nuclear energy technology in the CMEA, and its energy plans count on heavy dependence on nuclear power in the decades ahead. There is some evidence of the population's anxiety about the build up of nuclear plants; but while such worries could previously be easily dismissed as unjustified, after Chernobyl this will no longer be possible. Secondly, the regime is anxious that any manifestation of anti-Soviet emotions be nipped in the bud.[37]

On 7 May, the Charter 77 human rights movement responded to what it felt was a deplorable lack of information from the Czechoslovak authorities about the Chernobyl accident with a letter to the Federal Assembly and the Czechoslovak government. Noting that other governments, such as that of the Federal Republic of Germany, had issued warnings about the consumption of fresh milk, the Czechoslovak authorities had followed a terse statement that there had been no increase in radioactivity in the country (30 April) with five days of silence. The statement, signed in Prague on 6 May by Milos Palous, Anna Sabatova and Jan Stern, ended as follows:

Since the right to life and health is among the most fundamental human rights, we demand that you publish without delay all available information about the levels of radiation on the territory of the republic on each of the critical days. Of particular importance are sober estimates by experts who should inform the public about the risks still persisting and about the measures that should be taken now and in the future. At the same time, we demand that the government of the Czechoslovak republic request from the Soviet government all necessary information about the circumstances of the catastrophe [and that it] make this information public and let the public know what conclusions will be drawn from the Chernobyl accident, especially with a view to ensuring the safe operation of our own nuclear plants. Given the higher density of population in Central Europe, a similar event there could have even more far-reaching consequences. When it comes to the menace of radioactivity we must act in unison, because radioactivity recognizes no borders.[38]

The first notable counter-response of the government was a statement by Stanislav Havel, Chairman of Czechoslovakia's Atomic Energy Commission, on 14 May 1986, that the accident was "no reason" to alter the programme for nuclear power construction in the country. Havel declared that although "current" radiation levels in Czechoslovakia were one-and-a-half to two times higher than normal, a "very quick decline" was taking place. He also added that the safety of Czechoslovakia's nuclear plants was "comparable and certainly not lower" than those in the West.[39]

These reassuring comments evidently did not satisfy all citizens. Shortly after the Chernobyl disaster, a new group called "Anti-Atom" came into existence, and evidently began distributing postcards, with a photograph of a nuclear power plant on the front and a warning against constructing the Temelin plant on the reverse side. The slogan of the group is "Atoms against Life, Health and Nature." According to a Western source, the group also issued a strong statement protesting Czechoslovakia's plans for the development of nuclear energy, which noted that the reactors at Bohunice, Mochovce and Dukovany were not only located dangerously close to major population centres, but also did not possess twin containment structures to shield the radioactive zone.[40] The group appears to have a strong anti-war facet in its makeup, maintaining in its statement that through the build-up of nuclear plants, Czechoslovakia is becoming the nuclear base of the Warsaw Pact.

In addition to internal protests related to the Chernobyl accident, the Czechoslovak authorities also faced some external opposition in the shape of a group of Austrian students and academics, who launched balloons in Prague and distributed leaflets on Prague's Charles Bridge on 26 May, protesting against the construction of the Temelin nuclear plant. The aim was apparently to express Austrians' concern about the building of a giant plant so close to Austrian territory, and following the decision of the Austrian government to abandon its own nuclear programme before the Chernobyl disaster occurred. Five students were arrested, but subsequently released by the Czechoslovak police.[41]

The students' protests received formal backing from the Austrian government a week later, when Austrian Foreign Minister, Leopold Gratz, in a meeting with his Czechoslovak counterpart Bohuslav Chnoupek in Vienna, voiced his concern over the safety of "two Soviet-designed nuclear power plants operating...near the Austrian border," in addition to putting in request to the Czechoslakian government that it reconsider its plant to construct the nuclear plant at Temelin.[42] On the following day (3 June), however, Chnoupek appeared on Austrian television and announced that Czechoslovakia would continue its nuclear energy expansion despite the accident at Chernobyl, because nuclear power "is the energy of the future."[43]

The Austrian protests were not simply a formality. Although Austria has abandoned its own plans to build nuclear power plants and even dismantled its existing station, it concluded an agreement with Czechoslovakia in 1984 that stated that the two countries should exchange views at two-year intervals on "the development of nuclear programmes in individal countries." In fact, six months before a new reactor is to come on-stream, the agreement states, it is to be discussed by both countries. This occurred for the first two Dukovany reactors and a third meeting for the next power block was scheduled for June 1986. Although it is clearly under more pressure than hitherto to slow down its expansion, the Czechoslovak government cited this agreement as evidence of the sort of international co-operation which, in line with enhanced safety standards, will enable the future development of nuclear power in the country to continue at the same pace as before.[44]

In Poland, there were minor demonstrations in the southern city of Wroclaw as early as 2 May. A small protest, organized by the *Wolnosc i Pokoj* (Freedom and Peace movement), lasted for approximately forty minutes before a reportedly responsive crowd of 150 to 200 people. About eight people on Swidnicka Street carried posters with slogans such as "Zarnowiec Will Be Next," "We Demand Full Information" and "Is Nuclear Death from the East Different?" As the group protesting grew in size to twenty-two persons, the militia moved in in force and made some arrests.[45] Perhaps the most significant factor about the affair was the evidently warm response of a crowd to a demonstration so clearly directed against the future of nuclear power in Poland.

At a follow-up demonstration on 9 May, in which about fifty persons, predominantly mothers, took part, one of the demonstrators held a placard with the words "Today Chernobyl, Tomorrow Zarnowiec," showing that once again a direct link was being made between the Ukrainian accident and the Polish nuclear programme.[46] On 3 May, the Freedom and Peace Movement had issued a statement that demanded first that the Polish government should reveal more information about Chernobyl, and second that the authorities should cease construction work on the Zarnowiec plant.[47] The Yugoslav *Borba* newspaper also noted in its edition of 6 May 1986, that although a radiation cloud had been hanging over Poland for at least a week, only "today" [5 May] had the first specific data been published about the levels of radiation in the country between 28 April and 2 May. The newspaper stated that such incomplete and belated information had caused great indignation among the people and mistrust of official communiques.

On 15 May the Polish government issued a statement through its spokesman Jerzy Urban that the Chernobyl accident would not change its plans to build the country's first nuclear power plant with the aid of the USSR. The Zarnowiec construction was to continue and the first reactor would come on-line in 1990 (a year earlier than anticipated).[48] A more conciliatory approach, however, was adopted by the newspaper *Zycie Warszawy*, which noted in its front-page commentary of 16 May:

The age of the atom sets unusually high requirements for all, but especially for politicians. For not only nuclear energy is involved but also the security of all nations. Those having their own reactors as well as those suffering the effects of reactor breakdowns outside their borders. As soon as we realized the scale of what had happened, or what could have happened later, we saw how little can be done to counteract the consequences of a nuclear failure. World-scale co-operation is, perhaps, the most efficient method for making ourselves safe in future. Whatever the political system, it is necessary to act in the name of the supreme goal, which is life.

Polish news services also reported on 16 May that a meeting of the Polish Ministry of Energy had taken place and resolved to locate Poland's second nuclear power plant—called WARTA—in the Pila region of northwestern Poland.[49] Subsequently, the *Kurier Polski* newspaper divulged that the plant would be located near Szamotuly, on the Warta River, northwest of Poznan, and that it would consist of four 1,000 megawatt reactors, the first of which would be scheduled to come on-stream in 1995.[50] The initial announcement about the Warta plant on 16 May coincided with more anti-nuclear protests.

On this same day, over 3,000 people gathered in the city of Bialystok, which is located in the northeastern part of the country, i.e., the region most affected by the radiation cloud from Chernobyl. A petition was signed which asked tha Polish authorities to suspend construction at Zarnowiec, because it was felt that the plant's safety could not be guaranteed. The petitioners would sanction further work there only under the constant supervision of the International Atomic Energy Agency (IAEA). It was also revealed at this time that five prominent Polish scientists had addressed a letter to the government on 6 May, asking that it revise its dangerous decision on security systems at the plant.[51] Jerzy Urban made a further statement on 20 May that the government had decided to review the safety systems planned for the Zarnowiec plant, and would even consider safety equipment manufactured in the West. But at the same time he maintained that the safety standards at Zarnowiec were up to world standards and that Poland could not abandon its nuclear programme as it faced an energy shortage and did not wish to become "anachronistic and backward."[52]

The Freedom and Peace Movement organized another demonstration that took place in Cracow on 1 June, made up of about 2,000 people, according to Western reports. A news conference followed the protest, which ended at Wawel Castle and appears to have been motivated both by the radioactive fallout from Chernobyl and the arrest of the Solidarity underground activist Zbigniew Bujak. At the conference, the spokespersons demanded reparations from the USSR to the families of those killed in the accident, and criticized the lack of information provided by the Polish government.[53]

The response from the government was prompt. Jaruzelski stated that he understood the anxieties of Poles, but that there was ''no alternative'' to the continuation of work on Poland's first nuclear plant. He also declared that Poland had a political and moral obligation to support the USSR in its time of misfortune after the Chernobyl disaster.[54] By 20 June, the Commission set up following Chernobyl had prepared a report that envisaged changes in the design of the safety mechanisms at the Zarnowiec station, including the improvement of warning and measuring systems. Evidently the report contained many suggestions, and the Chairman of the Polish Nuclear Agency, Mieczyslaw Sowinski announced that it was to be published and discussed in the media.[55]

These events show that the Polish authorities felt obliged to make some response to the protests of their citizens even if by 16 June 1986, no decisive measures had been revealed. The likelihood is that the Zarnowiec plant will be subjected to further delays and consequently Poland will be more than ever reliant upon imports of electricity—mainly generated by nuclear plants—from the USSR.

In Bulgaria, which does not appear to have been affected by protests, there is nonetheless evidence that the authorities were somewhat nervous about the possible impact of Chernobyl. On 29 May, the official news service (BTA) included an interview with the country's most prominent nuclear scientist, Nikolai Todoriev, Chairman of Bulgaria's Energetika (power generation) and Corresponding Member of the Bulgarian Academy of Sciences. Todoriev appears to have had but one purpose: to assure listeners of the absolute safety of the Kozloduy plant. Clearly the scientist had no new information to impart. Instead he noted that the plant had been unaffected by an earthquake that had shaken Romania in 1977: ''At that time the nuclear power station experienced an earthquake of a higher scale than the one it was designed to withstand. Despite that it never stopped operating.''

Todoriev also remarked on how the Kozloduy plant was being constantly updated:

The entire equipment at the Kozloduy nuclear power plant was further updated after the devastating earthquake in Romania. The safety systems were improved and a number of new devices detecting the fluctuations of the earth's crust were fitted. The supplier of the power plant, the Soviet Union, is unceasingly working on its improvement. All new discoveries and innovations are rapidly introduced.

Even if an emission of a radioactive vapour occurred, Todoriev con-

67

tinued, it could not reach the outer environment, but would be contained in the reactor building.[56] The above statement's appealing and almost conciliatory tone was in contrast to earlier statements by the Bulgarian authorities, which concentrated on minimizing the impact of the Chernobyl accident in Bulgaria, and focused largely on nuclear accidents in Western countries, following the example of the Soviet news services.[57]

The Hungarian government made one concession to public concern over Chernobyl: a statement to the effect that the Paks plant reactors had better safety features than those at Chernobyl since they were within a containment building and used a double as opposed to a single water circulation system. An energy planning official, Robert Targan, also stated on 11 June 1986 that (like Poland) Hungary possessed no alternatives to nuclear energy because of the increasing domestic electricity demand. Those who fear nuclear energy, remarked Targan, "do not understand the technical details of Hungarian reactors." Rather than cut back on nuclear power, he informed, the country will triple the share of nuclear power in electricity production by the end of the century from 5 to 15 per cent.[58]

If the Hungarians thus far have been somewhat muted in their protests, the East Germans have clearly been more affected by the impact of Chernobyl, perhaps especially because of the influence of West German television in much of their country. According to West German reports, several hundred East Germans have begun a campaign to try to persuade their government to abandon nuclear energy by 1990. The government received a petition which reportedly requested detailed information for the public on the dangers of nuclear energy. It attacked the secretive practices in particular of the Soviet and East German governments. Again, the only definite response from the government in question—a response, moreover that preceded the delivery of this petition—was a statement that safety standards would be reviewed but that the nuclear programme would not be affected.

While most of the governments of the CMEA countries issued statements assuring citizens of the high safety standards of domestic nuclear plants, the most decisive actions occurred in Yugsolavia which, as shown, possessed a powerful anti-nuclear lobby before the Chernobyl accident. On 6 May, the Croatian Assembly drew up its development plan for the 1986–90 period and omitted completely any mention of the proposed nuclear plant construction at Prevlaka. It also pointedly left out any reference to the proposed site of a dump for low and medium radioactive waste from the Krsko nuclear station.[59]

What was described as the "indefinite postponement" of the Prevlaka plant was confirmed on the following day by the *Vecernje Novosti* newspaper, which quoted Grisos Curin, a delegate to the Croatian Assembly, as saying that the long-term plans for the Prevlaka plant should be revised following the Soviet mishap. Two days earlier, the *Politika Ekspres* had called for laws to stop the future development of nuclear power in Yugoslavia and to close down the Krsko plant. Public demonstrations followed.

On 10 May, for example, "several hundred young people" staged a demonstration in Ljubljana, Slovenia, and read a statement condemning the lack of information provided about Chernobyl. The protesters demanded that a request for compensation from the effects of radioactivity be placed before the Soviet government, and that a moratorium be declared on the construction of nuclear plants in Yugoslavia.[60] On this same day a meeting of Communist youth in Serbia also adopted a resolution "vigorously" opposing the construction of nuclear power plants in Yugoslavia, and a resolution on this question was reportedly adopted by the Eleventh Congress of Serbia's Federation of Socialist Youth. The latter declared that the building of nuclear plants would have serious ecological and economic consequences.[61]

There have been at least two other indications that opposition to nuclear energy has mounted in Yugoslavia. The leading political weekly *Nin* stated frankly that Chernobyl had destroyed the Prevlaka plans completely, and quoted an official as stating that "The fate of Prevlaka was decided in Kiev."[62] On 26 May, Yugoslavs putting on a gymnastics display in Belgrade to celebrate the birthday of the late Josip Tito sang an anti-nuclear song and formed an anti-nuclear sign on the field during the performance.[63] In contrast to the CMEA countries, official bodies have participated in anti-nuclear protests. The debate is being carried out at the top level, whereas in neighbouring states, governments anxious to pursue extensive nuclear programmes are facing reaction from below.

There is little evidence, however, to suggest that the Yugoslav Planinc government has abandoned completely its nuclear programme. Nevertheless, the programme has received a severe setback. A report in the British newspaper *The Guardian*, of 13 June 1986, suggested that the federal government had been obliged to "postpone indefinitely" its plans to develop nuclear energy in the wake of Chernobyl and that an opinion poll published in a Belgrade newspaper indicated that three out of four Yugoslavs were opposed to nuclear power. On 11 July, it was reported that power workers had delayed progress on construction of Yugoslavia's second nuclear power plant at Prevlaka by withholding (by vote) funds for feasibility studies. Consequently, building work will not begin until at least 1990.[64]

At the time of writing, however, no definitive steps had been taken and certainly no official statements had been issued rejecting nuclear power completely. Postponements and delays appear inevitable, but these may well have occurred without Chernobyl. Perhaps the main significance of the protests and demonstrations in Yugoslavia is their impact on a government that has hitherto played a positive and active role in the East European nuclear programme. In turn, through CMEA activities, the influence could spread to other countries, particularly those hit hard by the effects of the Ukrainian disaster.

On 19 May, a three-day CMEA Session was held in Moscow, at which time the USSR reportedly briefed the foreign delegations about the measures it had taken in response to the Chernobyl accident. Also the Session discussed the results of the "first stage" of the "scientific-technical" programme that had been adopted at the December 1985 Session, and dealt primarily with the nuclear energy programme.[65] The Session revealed that no changes had been made in the CMEA nuclear programme as a result of the Chernobyl accident. Not even a slight delay was anticipated. Thus according to a Soviet report, the agreed plan is that by 1 January 1991, the capacity of nuclear power plants in the CMEA countries is to be increased to a capacity of 35,000 megawatts, encompassing a rapid development of both nuclear power plants and nuclear power and heating plants. To assuage any doubts about the viability of such a programme, the Session reportedly declared that "the development of atomic energy...is the primary direction in resolving the energy problems in the socialist countries."[66] Three days later, *Pravda* confirmed that all the countries involved in the CMEA co-operation had decided to keep to their original programmes.[67]

The CMEA's decision, however, flies in the face of reality. In the past, even under the most favourable of circumstances, its participants have failed to maintain schedules for a programme that is conservative compared to the current ambitions. And yet after a setback that at the very least is a major psychological adversity, no provisions have been made for the expected slowdown in construction that will follow. Possibly the Soviets persuaded their partners to agree to their basic premise that Chernobyl was a setback, but not the deathknell for the nuclear programme. But the opposition to nuclear power in Eastern Europe that has already manifested itself in Poland, Czechoslovakia and Yugoslavia will make it almost impossible for these governments to keep on schedule. Instead, increased pressure will be placed on the leader of the programme, the USSR itself, and particularly its link-up region for East European plans, Soviet Ukraine.

CHAPTER FOUR

Ukraine in the Soviet Nuclear Energy Programme

The 1970s saw an ambitious expansion of nuclear energy in the USSR. At the end of the Tenth Five-Year Plan (1976–80), nine nuclear plants were in operation throughout the country, with a total of twenty-four reactors on stream. The locations were as follows: Leningrad, Chernobyl, Novovoronezh, Kursk, Kolsk, Beloiarsk, Rovno, Bilibino and the Armenian republic. Graphite-moderated reactors, like the one at Chernobyl, predominated, with thirteen reactors accounting for 58.6 per cent of total electricity capacity; in addition there were ten light-water pressurized reactors (VVERs), which at this time was a new experimental reactor first brought into operation at the Novovoronezh nuclear plant, and one so-called fast breeder reactor at the Bilibino nuclear heat and power station. Most of the reactors in operation were of 1,000 megawatt capacity.

With a total capacity of 12,500 megawatts, the nuclear plants made up about 5.6 per cent of the total electricity production in the USSR, and 9.3 per cent in the European part of the country,[1] where most of the plants were located.

In terms of overall electricity production in the USSR, nuclear energy has lagged well behind thermal and hydro-electric power stations. In 1980, of the total electricity output of 1293.9 billion kilowatt hours, thermal electric stations accounted for 1037.1 billion kilowatt hours (80.2 per cent), hydro-electric stations for 183.9 (14.2 per cent) and nuclear power plants for only 72.9 billion kilowatt hours of production. The Eleventh Five-Year Plan for the 1981–5 period, however, anticipated a rapid increase in the output and proportion of nuclear energy in electricity supply by approximately 300 per cent. The overall share of thermal electric stations was expected to decline to 71.1 per cent by 1985, hydro-electric stations were to rise slightly to 14.8 per cent of

71

the total, while the share of nuclear power was to rise suddenly from 5.6 to 14.1 per cent, and in the European part of the USSR, from 9.3 to a significant 23.8 per cent by the end of 1985.[2]

The Eleventh Five-Year Plan was somewhat ambitious in scope and proved to be well beyond the capabilities of those constructing Soviet nuclear power plants. It did, however, mark the beginning of what was called the first stage of the nuclear energy programme that is to continue into the twenty-first century. By 1985, it was envisaged that the 12,500 megawatts of capacity at nuclear power plants would have been raised to 33,800 megawatts in the USSR, and that 98 per cent of the European part of the country's increased electricity needs would be met by this source. In short, while nuclear energy was not yet to be the dominant force in Soviet energy, it was the sphere designated to meet the heightened demand for electricity.

The expansion took two forms. First, capacities were to be raised at already existing stations so that by 1985, three of the sixteen plants expected to be on-stream would be over 4,000 megawatts in size, with another six plants between 2,000 and 3,000 megawatts in capacity. The plants in question included those at Kolsk, Kursk, Chernobyl, Leningrad and Rovno. In addition, nine new nuclear plants were to come into service, all of which had been under construction before 1 January 1981: Kalinin, Zaporizhzhia, Rostov, Balakovo, Khmelnytsky, Crimea, Ignalinsk, Smolensk and South Ukraine. In geographical terms, four of the nine were in Soviet Ukraine, four in the Russian Republic and one in Lithuania (Ignalinsk). With the exception of Smolensk and Ignalinsk, the new reactors to be constructed were of the 1,000 megawatt water-pressurized variety.[3] The rate of expansion of the graphite-moderated reactors was thus to be decreased. Altogether a further twenty plants were under construction, including four nuclear power and heating plants and two nuclear heat supply stations, all in the European part of the country.

From the outset, there were problems in adhering to the plan. In 1981, for example, only three out of eight reactors scheduled actually came into operation, and a further six were said to be behind schedule.[4] Nevertheless, the increases in capacity were considerable. In 1981–2, for example, the capacity of nuclear power plants in the USSR rose by 38 per cent.[5] In 1984, according to Soviet accounts, total capacity increased from 17,000 megawatts in May,[6] to 21,000 in December, at which time the proportion of nuclear energy in electricity output was said to be just under 10 per cent,[7] a major rise although well short of the proportion anticipated.

By December 1984, the Leningrad nuclear power plant had reached a capacity of 4,000 megawatts, accounting alone for 2 per cent of the USSR's supply of electricity. Chernobyl had reached this same size by

1983, while 1984 also saw the first output at the Zaporizhzhia station in the industrial heartland of eastern Ukraine.[8] According to a Western source, however, the 1984 plan for the installation of nuclear capacity fell short of its target of 5,400 megawatts by 2,000 megawatts, and it was these sort of delays that accounted for electricity output missing its goals for 1984.[9]

The plan for the 1985 year was to bring into exploitation energy "blocks" of 1,000 megawatt capacity at the Smolensk, Kursk, Balakovo and Zaporizhzhia nuclear plants, according to *Radio Moscow* (1 January 1985). It was reportedly carried out on schedule, although not without a major struggle at the Balakovo station, which came on stream only on 25 December following a multitude of problems. The fourth unit at the Kursk station also came into operation at the very end of the Eleventh Five-Year Plan.[10] One's assessment of the results of the 1981–5 Plan must take into account two factors: first that a remarkable increase in the capacity of nuclear plants was attained, from 12,500 to over 25,000 megawatts, or double the total at the end of the previous five-year plan. Second, that in terms of fulfillment, by the standards that the Soviets have set themselves, the plan was an abject failure. The final total did not even approach the anticipated 33,800 megawatts, making up only about 74 per cent of the targeted figure. In the summer of 1985, the longstanding Minister for Power and Electrification of the USSR, P. Neporozhny, was removed from his post, although this may have had as much to do with problems in the oil industry as with nuclear energy. It is clear nonetheless that like other countries of the world, the Soviets had found it difficult to adhere to a timetable in the development of nuclear power, even in a situation in which there was no internal opposition to the industry's development and although the Soviet system's centralized makeup enabled resources to be poured into a zone of industrial expansion.

Brezhnev's mantle was taken up by Mikhail Gorbachev, who appears to be in full agreement with his predecessors about the wisdom of expanding nuclear energy to unprecedented proportions, in a bid to bring about the so-called "scientific-technological acceleration" of the Soviet Union by the year 2000. Indeed, according to the First Deputy Chief of the USSR State Planning Committee, L. Bibin:

In electrical energy, one of the key directions of scientific-technical progress is to raise the proportion of electricity generated at atomic energy stations. The output of these stations in the current plan [1986–90] is to be almost doubled compared to 1985....There will come into operation the second and succeeding energy blocks at Kalinin, Balakovo, Zaporozhye [Zaporizhzhia], Rovno and Chernobyl atomic energy stations and the first block at the Khmelnitsky [Khmelnytsky] AES.[11]

In fact, according to the accounts published of the proceedings of the Twenty-Seventh CPSU Party Congress of March 1986, the Twelfth Five-Year Plan is even more ambitious than Bibin stated. Total electricity production is to be raised from the 1985 output of 1,545 billion kilowatt hours to 1,840–1,880 billion kilowatt hours by 1990, although according to Minister of Power and Electrification of the USSR, A. Maiorets, this figure is perceived as the "lowest possible target."[12] Of this total, about 390 billion kilowatt hours is to be made up by nuclear power plants, which accounted for only 170 billion kilowatt hours of output in 1985.[13] In terms of capacity, the scale of the build-up becomes more comprehensible with Maiorets' statement that whereas the total capacity of the nuclear power plants in the USSR stood at 28,000 megawatts in March 1986, it is to compose 69,000 megawatts by the year 1990, or an increase in the region of 250 per cent.[14]

This expansion can be compared in scale only with the earliest Stalin five-year plans, and in terms of actual implementation, it is unlikely that it could be fulfilled without major changes in the way Soviet industry operates. For the purposes of this study, however, there are two factors of major significance in the new programme: the first is that the planned expansion is to be largely at existing rather than new sites. Thus the average size of the nuclear power plants in the USSR is to be enlarged considerably. Two staff members of the Moscow-based Institute of Nuclear Power wrote in November 1985 that the authorities intend to build in the future mainly big nuclear plants with a capacity of 4,000–7,000 megawatts.[15] The Twelfth Five-Year Plan also anticipates the construction of "huge atomic power stations with a capacity of 4,000–6,000 megawatts,"[16] while in the autumn of 1984, *Izvestiia* stated that the average capacity of Soviet nuclear power plants in the 1986–90 plan period is to be 6,000–7,000 megawatts.[17] While there are clearly some advantages to expanding existing nuclear reactor sites rather than building new ones, such as having a workforce already in place and social and community buildings available in addition to housing, there are innumerable incumbent problems, particularly regarding safety factors in building such massive nuclear power plants.

Second, the planners intended to get the programme under way at a rapid pace in 1986. At least six new reactors were to come on-line in 1986, and even though some of these were well advanced at the time of the formulation of the plan, in other cases timetables were brought forward to achieve this new goal (for example, at Chernobyl itself). As with other sectors of the Soviet economy at various times, nuclear energy had been made the subject of "shock-work." A vast programme had been undertaken, in spite of the fact that much less ambitious plans had failed in the past, and regardless of the nature of an industry that demands considerably more attention to safety and other factors than most.

The Ukrainian SSR is playing a pivotal role in the expansion of nuclear energy in the Soviet Union, partly because of its location in the industrial heartland of the European part of the country, and partly because its proximity to East European countries enables it to participate in the CMEA programme encompassed within the MIR system. Compared to the Russian Republic, in which a small reactor at Obninsk was operational as early as 1954, nuclear power in Ukraine is very much a phenomenon of the 1970s and 1980s. Construction of the first nuclear power plant—Chernobyl—began in 1969–70,[18] while in 1974 it was hinted in the Ukrainian press that experts from the Kiev Institute for Industrial Energy Planning (*Promenerhoproekt*) were to construct the USSR's first nuclear power and heating plant near the city of Rovno in Western Ukraine.[19] (In fact, Rovno was actually built as a conventional nuclear power plant.)

The Tenth Five-Year Plan (1976–80) foresaw a total of 4,880 megawatts of capacity at Ukrainian nuclear power plants, at which time three stations were to come on-stream at Chernobyl, Rovno and South Ukraine (two each). In the event Ukraine failed to meet these targets and in 1980, only the Chernobyl and Rovno stations were operational, with an aggregate capacity of 2,440 megawatts, i.e., exactly 50 per cent of the planned requirements. The 14 billion kilowatt hours generated by the Rovno and Chernobyl stations accounted for a mere 6 per cent of total power production in Ukraine, but about 19 per cent of nuclear power output in the USSR as a whole.[20] Thus while Ukraine had fallen well behind production schedule, it was already playing a significant role in the Soviet nuclear energy programme.

The Eleventh Five-Year Plan forecast a total nuclear capacity in Ukraine of almost 10,000 megawatts by 1985. The Chernobyl station was to complete the construction of two more reactors for a total capacity of 4,000 megawatts, a second 440 megawatt reactor was to come on-line at Rovno and the delayed South Ukrainian station was to complete work on its first reactor and bring a second into service before the end of the plan. In addition, new plants were under construction at Zaporizhzhia, Khmelnytsky and the Crimea, all of which were to be in service by 1985. In short, the proportion of Ukrainian nuclear plants in the overall Soviet total was to rise from the noted 19 per cent to 29.2 per cent over the course of the plan.

The Plan enjoyed mixed success. At Zaporizhzhia two reactors came on-stream in short order, but the Crimean and Khmelnytsky stations fell well behind schedule and were not in service by the end of the plan. Ukraine's total capacity in 1985 was 8,880 megawatts,[21] 90 per cent plan fulfillment, but over 35 per cent of Soviet capacity. After the Russian Republic, Ukraine was now the principal area for nuclear power development. The general outlook in Ukraine was discussed by the Ukrainian Minister for Power and Electrification, V. Skliarov, in February 1984.

Skliarov noted that seven nuclear power stations were being constructed on Ukrainian territory: Chernobyl, Rovno, South Ukraine, Zaporizhzhia, Khmelnytsky, Crimea and Odessa (a nuclear power and heating station). The first three were already in operation (as noted), and Zaporizhzhia was progressing "at a rapid tempo," as the main station in a series being built according to a unified pattern on the basis of VVER 1000 (water-pressurized) reactors. The CMEA countries, he stated, and principally Poland, were taking part in the building of the Khmelnytsky plant, which is to have a total capacity of 4,000 megawatts, based on four reactors. Skliarov also revealed that something had gone seriously wrong with the construction at Crimea, which was now scheduled for the Twelfth Five-Year Plan (1986–90), even though originally its completion had been foreseen for the 1981–5 plan period. Skliarov also emphasized the importance of the 750-kilovolt transmission lines that are linking nuclear plants with electricity consumers. He stated that a line connecting Chernobyl with Vinnytsia oblast was already functioning and that there would soon be others connecting the South Ukrainian station with Moldavia and the Zaporizhzhia station with the Dniprovsky and Zaporizhzhia substations.[22]

In the summer of 1985, however, Borys Kachura, a Secretary of the Politburo of the Communist Party of Ukraine, stated that because of the lagging permitted in the construction of the Rovno and Khmelnytsky nuclear power plants, the amount of energy capacity in operation in the Ukrainian SSR during the Eleventh Five-Year Plan was 2,000 megawatts less than anticipated. To improve this situation, he continued, it was necessary to make use of the report of the energy construction leaders at Zaporizhzhia, which concerned the application of high-speed construction methods, based on the unified flowline system. There, as a result of the assembly of large blocks in factories beforehand, new generating units were being brought on-line at intervals of one year, or, at the most, eighteen months. Kachura also noted that over 26 per cent of the transmission lines in Ukraine were in a state of disrepair, and the Ministry of Energy of the Ukrainian SSR "must explain how the condition of the electro-transmission lines can be improved."[23]

Skliarov and his colleagues, like their all-Union counterparts, were evidently undaunted by Kachura's remarks or by the slow progress at many Ukrainian nuclear plant constructions, and made very ambitious plans for the 1986–90 period, motivated, as shown above, by the stagnation of Ukraine's all-important coal industry and its need to rely on the equally unreliable Siberian coal. Now nuclear power was to make up this shortfall. By 1990, nuclear energy was to account for almost 40 per cent of Ukrainian electricity production,[24] i.e., considerably higher than the Soviet average.

By 1990, Ukraine's task is to bring its electricity production up to 320 billion kilowatt hours by completing the construction of the nuclear power plants at Chernobyl, the Crimea, Zaporizhzhia and Odessa, and bringing either first or new energy blocks into service at Khmelnytsky, Rovno and South Ukraine.[25] The biggest tasks, however, were assigned for the 1986 year, during which time the Ukrainian nuclear plant capacity was to be raised by 150 per cent (before the Chernobyl accident rendered such an increase impossible). In detail, the 1986 year was to see the following reactors brought into operation: number three at Zaporizhzhia; number five at Chernobyl; number three at Rovno, which was to be of the 1,000 megawatt variety as opposed to the two earlier 440 megawatt reactors at this station; and number one at Khmelnytsky. Major advances were also be made on the constructions at South Ukraine, the Crimea and the Odessa nuclear power and heating plant.[26] In addition, construction was scheduled to begin on a second Ukrainian nuclear power and heating plant to service the city of Kharkiv.[27] Had the annual plan been met, Ukraine's nuclear capacity would have risen from 8,880 to 12,880 megawatts in the space of a single year, which would have been by far the greatest annual increase in the history of the Soviet nuclear power industry.

These figures illustrate the sheer size of the Soviet programme, and demonstrate that Ukraine was to become the major area of expansion. Before dealing with the Ukrainian nuclear power plants in more detail, it should be emphasized that the overall plan for Ukraine had created great pressure upon the workforce. Nuclear energy is a relatively new industry. It is not like the coal industry where, for example, if easily accessible reserves were available, a build-up of this nature would be conceivable because the skilled workforce and necessary infrastructure already exist. And while one should not decry the level of skills manifested in various spheres by Soviet scientists and Soviet engineers, Soviet industry as a whole is dogged by a variety of problems, which might be divided into a shortage of skilled labour, difficulties in obtaining raw materials and problems in organization, related largely to the centralization of planning.

In 1984, when the first reactor at the Zaporizhzhia nuclear power plant came on-line only four years after construction began, the Soviet authorities declared that it was to be a model for the future, and that serial production at nuclear plants had proved to be viable. In brief, the Zaporizhzhia example was taken as evidence that rather than construct nuclear plants individually, based on local needs and the requirements of, say, a specific region, serial production, in which major sections of a power block are produced in factories, was possible. As Skliarov declared:

> Nuclear power plants are being improved. A unified nuclear project has beewn created whereby nuclear power plants, whose power blocks have a generating capacity of 1,000 megawatts, would be built in factories using large assembly pieces in their construction....The Zaporizhzhia atomic energy station, with an ultimate generating capacity of 6,000 megawatts, is being built rapidly via this method of construction.[28]

All the indices and restraints of a centralized planning system could therefore be applied to nuclear energy as to other industries. In some respects, there were advantages to be gained by so-called "serial production" of the various components of nuclear power plants: for example, turbine production could be regularized and even the type of the reactors themselves could be made consistent.

Thus after Chernobyl nuclear power plant came into operation using graphite-moderated reactors, this type of reactor was no longer manufactured for Ukrainian plants. Instead, reactors at Ukrainian plants were made consistent with each other, and with those being made for use in East European plants, i.e., of the VVER 1000 variety. In this way, construction workers and engineers could become accustomed, for instance, to working with one specific reactor and one type of turbine, and can plan accordingly. The Soviets have described this form of production as the "unified flowline system" and maintain that it is because such a system has been introduced that the whole construction process can be speeded up considerably.

At the same time, however, by putting nuclear power on this basis, and by introducing a flowline system that relies largely on "shockwork" conditions—speed being of the essence—the authorities have heightened already existing difficulties with supply and skilled labour. Several ministries are supplying the nuclear power plants and the centralized planning system is proving cumbersome and unreliable in coordinating the supply of materials from various quarters. As for the quality of labour, which is of such vital significance in the nuclear industry, the expansion of this sphere inevitably means that the high-quality labour is dispersed over a wider area. In Ukraine, as will be shown below, as each new nuclear plant is tabled for construction work, the best and most experienced engineers arrive from other Ukrainian plants to supervise the building. Yet construction at the engineers' former plants does not cease: on the contrary, for the most part, it expands at an even greater rate, as is witnessed by the examples of Chernobyl and Zaporizhzhia. Consequently, at the time of most rapid acceleration of some of Ukraine's nuclear plants, the skill of the workforce at the site is at a lower level than hitherto. Simultaneously, the timetables for completion of an energy block are being reduced throughout the republic. Because of the shortage of labour, advertisements are run in newspapers throughout the Soviet Union—in Central Asia, for example, where notices have pointed out that "no experience is necessary" for Tadzhiks who would be willing to work at Ukrainian nuclear power plants.[29] A leading role in the construction of Ukrainian stations is being played by the Komsomol. Many young people are spending summer vacations working on constructions at nuclear sites. The end-result is a dangerously low level of skill and qualifications among some members of the workforce. This factor has added to some severe problems that were already affecting Ukrainian nuclear plants, at Chernobyl and elsewhere.

In looking at the construction of Ukraine's stations in more detail, it should be noted that the Chernobyl plant will be dealt with separately in Chapter 6. The following survey looks at some of the dilemmas encountered by other Ukrainian plants in the face of the expansion of nuclear power in the 1980s, and Ukraine's dominant role in that process.

Rovno Nuclear Power Plant

The Rovno nuclear power plant is located in the new city of Kuznetsovsk, about 80 kilometres north of the city of Rovno and an even shorter distance from the Belorussian border. Sometimes referred to as the Western Ukrainian station, its commercial operation dates from 1979, two years after Ukraine's first reactor was started up at Chernobyl, and a second 440 megawatt reactor came on stream during the 1981–5 Plan.[30] According to a Soviet source, the first reactor block was constructed in half the time permitted by the plan. Three thousand workers and 130 brigades were involved in the work, and the project leaders were said to have considerable experience in building "traditional power plants" and to represent many different Soviet nationality groups.[31]

At least some of those working at the Rovno site in the early years of construction came directly from work at the Chernobyl station, which at that time was the only other nuclear plant under construction in Ukraine. Moreover, the majority of workers there were said to be young, with either a small family or no family at all. As a result, in the atomic city under construction, 70 per cent of the apartments being built were either one or two-room facilities.[32]

Recent Soviet reports suggest that the Rovno nuclear plant is experiencing some severe problems in the building of the third (larger) reactor.

A Soviet newspaper complained in December 1985 that only slightly over half the workers were using the "progressive brigade system" in construction.[33] In February 1986, *Radio Kiev* announced that the new energy block at Rovno "must be generating" electricity in the first half of 1986, but said that over 25 million roubles assigned for the work had not been used, and that the entire collective of the Rovno section of the "Chernobyl energy" organization, which is carrying out the construction, was "practically without work."[34]

This report also noted that an "enormous volume" of work still had to be fulfilled, including 7,500 metres of heat isolation and over 12,000 square metres of metal covering. The head of the Rovno section, S.M. Shevchuk, stated that the work fulfillment rate stood at 20–30 per cent, and "I have to say that the quality of the work is very low." He went on to say that he had visited the machine room, but that "no stabilization had occurred there." The chief engineer referred to "planning overruns" necessitating the preparation of technological equipment that went well beyond the original cost estimates for the job. In 1985, 8,000 man-days were lost at the Rovno nuclear power plant, many of which were attributed to "violations of discipline." More specifically, many of the workers were said to be regularly late for their shifts and an increasing number had had to be sent out to "drying-out institutes" to try to cure them of alcoholism.[35]

Despite these immense labour problems, the Rovno construction workers' task was to bring the third reactor on-line by May 1986. Moreover, a substantial amount of building remained to be carried out in the atomic city of Kuznetsovsk, including hotels, heat combines, two pioneer camps, five kindergartens, two schools, a library, a "palace of pioneers" and a sporting complex, according to the Second Secretary of Rovno oblast party committee, I.I. Zahorulko.[36] Not surprisingly, the Ukrainian Minister for Power and Electrification, Vitalii Skliarov, revealed in June 1986 that the generation of electricity at Rovno's third reactor had been delayed until the autumn of 1986.[37]

South Ukraine Nuclear Power Plant

The South Ukraine atomic energy station was sanctioned by the Twenty-Fifth Congress of the CPSU in 1975, and construction got under way in September 1976.[38] It is located on the South Buh River near Prybuzhzhia (formerly Akmechetka), in Domaniv raion, Mykolaiv oblast. A new town, Konstantynivka, has been established close to the plant site. In 1976, it was envisaged that a three-year period would be needed before the plant came into operation. Initially the workforce was composed of people who had been working on the huge thermal electric station at Kryvorizhzhia, of whom about 250 arrived at the site of the South Ukraine station early in 1976. By the summer of this year, the size of the workforce had risen to over 1,000. The proposed plan foresaw the construction here not only of the nuclear plant, but, using the extensive natural river systems, of a huge energy complex that also included a hydro-electric power station.[39]

In the mid-1970s, there was strong evidence that problems had developed that made the future generation of the station in 1979 improbable. The head of construction work, O. Sosedenko, complained, for example, of a shortage of some materials, and that the plant was being built before roads had been laid and a railway network established (the latter would clearly have been essential for the supply of the various materials to the plant site). Sosedenko also pointed out that he had a very youthful workforce that was made up predominantly of young Komsomol members, over half of whom were unmarried.[40]

In 1977, there were further reports about the lack of skill of the builders of the plant and their need for further training in their trade. Supplies from factories were reported to be ''inconsistent.'' Perhaps even more serious, the then Ukrainian Minister for Power and Electrification, O.N.Makukhin, reported that in the period 1976–7, 24 reinforced blocks and 15,000 tons of reinforced concrete had already been used in the plant's construction. This huge amount of materials was said to be essential as the plant ''was being built in a zone of seismic activity.''[41] The river systems that proved so convenient for the location of this station, therefore, were to some extent offset by the natural geological drawbacks of building a plant in a seismic region.

The leaders of the construction of the South Ukraine seem to have either had previous experience working at thermal or hydroelectric power stations, or else came directly from other nuclear power plants. Thus, a foreman, P.I. Vereshchaka, had been a brigade leader at the Smolensk nuclear power plant, and his brigade had been at the forefront in the ''socialist competition'' there. In short, the man had built himself a reputation for getting the job done quickly. Another foreman, P.M. Speka, had worked on thermal electric power projects at Ladzhinsky and Zaporizhzhia before moving on to South Ukraine.[42] Similarly, construction work was under the supervision of T.I. Antifeev, who had been working previously at the Kolsk nuclear power plant.[43]

In August 1977, there were more official complaints from plant leaders about deficiencies in their work, which, taken together, give the impression of a disorderly team struggling to meet one problem after another, while clinging grimly to a rigorous timetable for work completion. One foreman, M.M. Hanzhela, complained that the supply problems

81

were slowing down the tempo of work, and that they were awaiting the arrival of sulphite cement. Another official referred to serious shortfalls in supplies. For example, only 97 tonnes of material had been delivered from the experimental structure factory in Kiev instead of the 450 tonnes ordered; 530 instead of 900 tonnes from this same factory during the first six months, and 2,500 cubic metres of wood material as opposed to the 9,000 required. Further, some of the materials arriving at the site were said to be substandard.[44]

An electrician who was described in the Soviet press as a "well known figure" at the South Ukraine station, S.M. Kolesnyk, with ample experience in construction work, alluded to the "initial chaos" at the plant site and the problems encountered with drills and improperly laid pylons. Kolesnyk said that many of the workers were unfamiliar with some of the details of the installation of electricity supply.[45] An earlier account had referred to the difficulties that some of the youthful brigade leaders had run into,[46] and clearly many of these early troubles at the site were a result of the inexperience of the workforce there.

By 1978, it was clear that the schedule for completion of the first reactor had been revised. According to a Ukrainian source from the summer of this year, the reactor was to have been ready by 1980, and the ultimate capacity was to be 6,100 megawatts.[47] Yet even this timetable proved to be over-optimistic and only in December 1982 did the first reactor become operational.

It should be noted that problems that might have been associated with the plant's early years continued into the eighties. A report of September 1981, for example, spoke of serious miscalculations in the plans for construction and the interruption of the supply mechanisms,[48] while in the following year, the weekly *Ekonomicheskaia gazeta* chided the USSR Ministry of Power and Electrification for problems in organization, planning, work quality and labour morale at both South Ukraine and Smolensk nuclear power plants.[49] The station has been plagued with building problems from the outset and these have not been diminished with the onset of time.

In the late 1970s, more details about the South Ukraine station emerged that revealed it to be a plant that was quite distinct from other Soviet nuclear plants. A series of dams were planned and reservoirs were being established at three different levels in order that the water might service not only the nuclear and hydroelectric stations, but also be of use in irrigating local farmland. The dams also helped by raising the water level so that ships could gain access to the nuclear site (presumably conveying necessary materials). Water was being brought to the energy complex from Lake Konstiantyniv (whence the nuclear town derived its name), and pumped through tunnels to Tashlytskyi Reservoir, which was the atomic station's cooling source. In 1978, a third power complex was being planned for the system, namely a joint hydro-electric and hydro accumulation station, which was to be located on the Konstiantynivka dam.[50]

As the power complex on the South Buh developed, the town was expected to expand in size from 40,000 in the late 1970s to an eventual population of 180,000, at which time it was anticipated that a palace of culture, a music school and a fully equipped hospital would be available.[51] In 1981, the situation became further complicated when the second reactor of the South Ukraine station, which like the first was to be a 1,000 megawatt water-pressurized type, was designated for Romanian and Bulgarian electricity requirements. In 1981, in Moscow, the Soviet Trade Minister N. Patolichev and his Romanian counterpart, C. Burtica, signed an agreement by which Romania was to receive 1200 million kilowatt hours of electricity in 1985, 2,400 in 1986 and 500 million kilowatt hours starting in 1988.[52] The agreement specified that power from the nuclear power plant was to be transmitted to Romania via a 750-kilovolt line that ran through Moldavia and across the Danube into Romania. In November 1985, this line was said to be close to completion, and to be a joint project of the Soviet Union, Romania and Bulgaria.

The report made clear that Romania's stake in electricity from South Ukraine was higher than that of Bulgaria, which indicates that Romanian investment in the project is also higher, since traditionally the amount of electricity received from Soviet power plants is directly proportional to the amount of funds invested by a given country. In the case of a breakdown at a power station in any of these three countries, the current was to be supplied by another transmission line from one of the other CMEA countries, where similar lines are being built.[53] Subsequently, *Radio Moscow* reported that the electric transmission line (described as an "energy bridge") from South Ukraine to Eastern Europe had been completed in March 1986, and that in addition to the above countries cited, Hungary, Poland and Czechoslovakia were also connected with this grid.[54]

Following the generation of the first reactor at South Ukraine in December 1982, the second came on-stream in March 1985.[55] At least one more block is planned for the 1986–90 Plan, and it is probable that two more reactors have been scheduled for this time period. The station is expected to be completed during the Thirteenth Five-year Plan of 1991–5.

Zaporizhzhia Nuclear Power Plant

The Zaporizhzhia nuclear power plant has been designated ultimately to be the largest station in Ukraine with a capacity of 7,000 megawatts. One reason for this is that it has been set aside as a model station, and the atomic city of Enerhodar is being established as a major centre not only for the training of skilled personnel in the nuclear industry, but even as a sort of exhibition area to be visited by representatives of the nuclear industry from friendly countries. Thus in June 1985, it was reported that the city had been visited by "specialists" from Bulgaria, Czechoslovakia and Mongolia, and that earlier in this year a group of Cubans had been the guests, studying under the supervision of engineers from the Kiev Institute of Construction Engineering[56] (Cuba is building a nuclear power plant based on water-pressurized reactors, with Soviet help).

As noted above, the Zaporizhzhia station is being constructed according to a standardized design that is said to enable several power blocks to be constructed in quick succession, ideally at yearly intervals. Through this system, the equipment is reportedly assembled in units that can be delivered to installation sites "in a state of maximum factory readiness."[57] This method of production is said to have resulted in a substantial reduction of labour costs and fuel, mainly by accelerating the whole process of building. Certainly there have been no references in Soviet works to improvements in quality that might arise from this system: the emphasis is on time-saving, first and foremost.

The Zaporizhzhia station had two reactors in operation by July 1985, with the second energy block coming on-line seven months after the first. Since the station is serving as a model not only for Ukraine, but also for the Soviet Union and East European countries, it is imperative that the hectic schedule be maintained. As a result, and although it appears on paper to be highly implausible, the timetable foresees the completion of all seven reactors here by 1990, with the third coming on-line in 1986, and the others scheduled to follow at intervals of one year.[58] The situation at Zaporizhzhia was highlighted in *Izvestiia* in April 1986:

> Nuclear energy has almost never known the tempo of construction that has been achieved at the Zaporozhye [Zaporizhzhia] AES. Recently, the first energy block, with a capacity of one million kilowatts came on-stream, followed by the second. And today, we are completing the construction of the foundation structure under the sixth [reactor]. The interval separating the coming on-stream of energy blocks has been reduced to a year! The basis of this high tempo is the unified flowline system of atomic energy stations, which was first introduced at Enerhodar.[59]

Does this suggest that Zaporizhzhia is free from the sort of niggling and more serious difficulties that have hindered the construction of nuclear plants in the USSR? The evidence indicates otherwise. On 28–30 October, the Ukrainian First Party Secretary, Volodymyr Shcherbytsky, paid a personal visit to Zaporizhzhia nuclear power plant, his first reported sojourn at a nuclear plant anywhere in Ukraine. It is not implausible that he felt it incumbent to visit such a model specimen during a routine that saw him pass through other industries of this oblast. But on the other hand, party leaders do not visit places by chance, and Shcherbytsky clearly had to go out of his way by at least 30–40 kilometres to make his stop. Moreover, his comments indicate that he had some serious reservations about the work being carried out there.

The Ukrainian leader spoke to plant personnel, focusing on some of the difficulties that are said to be delaying the third 1,000 megawatt power block. He noted, in particular, hold-ups in the delivery of nuclear machinery to the station, and the failure of the planning organs to provide a timetable for the acquisition of the equipment needed, and for the recruitment of the necessary personnel. He also revealed that the construction work was preceding the provision of workers' facilities, noting the absence of cultural and sporting facilities at Enerhodar, and he reprimanded verbally the city of Enerhodar and the raion authorities for their failure to improve transport, medical and commercial facilities.[60] In short, then, Zaporizhzhia, as a model specimen, shows that no Soviet plant has been free from these problems.

Odessa Nuclear Power and Heating Station

At Odessa, the problems appear to be more serious than at most other Ukrainian stations. Construction of the USSR's first nuclear power and heating plant began there in May 1981, and two 1000 megawatt water-pressurized reactors are being built to provide electricity and heating for the city of Odessa, which has about 1.2 million residents. At peak capacity, each reactor is expected to produce 90 kilowatts of electricity and 900 gigacalories per hour of heat.[61] A new town, Teplodar, is being established at the reactor site, and Soviet reports have frequently referred to the Odessa station, with pride, as the forerunner of many such nuclear power plants designed to meet the heating needs of major Soviet cities. In Ukraine, for example, similar plants are to be built in Kharkiv and at Kiev itself.

That major setbacks had occurred at Teplodar became evident from a Soviet Ukrainian press report of October 1984, which discussed the "slow construction of the industrial base at the Odessa ATETs."[62] The report went into unusual detail and as such merits an in-depth analysis. It began by depicting a quarrel between the firechief at Odessa station, A.I. Bener, and the chief engineer of the construction organization, M.V. Feshchenko. The fire chief apparently had been hanging on to the assembly details for the plant for over three weeks because he felt that the way the station was being constructed constituted a fire hazard. Bener was quoted as saying that although 4.5 million roubles had been spent on fire protection, there were still no "anti-fire conductors," "no protection from lightning" and not even the "primary means" to put out a fire. The roof (presumably that of the building housing the reactor) had been made "anyhow" and the passages needed for protection from fire had been blocked up. It should be added that the tone of the article implied that Bener was taking an unreasonable stance and that Feshchenko had been put in a humiliating position of having, more or less, to beg for the assembly details he required.

The article then focused on the tardiness in the building of the plant, which was attributed to both lack of organization and generally sloppy work. It was stated that in the period since construction began in May 1981, the builders' colleagues had already constructed a naval berth at Bilhorod-Dnistrovsky and a network of roads, but "the builders [of the atomic plant] often do not comprehend the costs of the time spent on preparation." A brigadier at the site, A.V. Plaksienko, noted that "I have worked on constructions for almost three decades, but never have I found such a lack of co-ordination as here." One wall of concrete, he added, had had to be laid three times because the junctions were not at the proper levels. Again and again, the cement mixture provided by the Odessa section of the Institute for the Organization of Energy Construction was unsatisfactory.

As for the actual process, the way in which the cement mixture was made, "no one understands" how it works, and the workers had to wait for the arrival of either the chief engineer of the plan or an institute assistant before any decision could be made. One official declared that in August 1984 alone, he made thirty statements about alterations and extra work that had to be carried out. The implication is that the workforce simply lacked the experience for the task. The concrete was made up of sand and crushed stone filler, but the whole area from where it was obtained was flooded with groundwater and "work has stopped completely" because it was not possible to assemble the transporters and other equipment in the flooded zone.

In any case, the concrete gallery constructed by the sections of *prombud* [industrial construction] of the station had been badly built, while the internal isolation work was said to be of low quality. Despite this chaos—for the whole operation appears at this point to have been a fiasco—the man in charge of the entire construction, Volodymyr Dubensky, was said to regard the situation "with indifference." The report stated that he had no fewer than fifty-eight specialists under his leadership, but that the building was lacking in co-ordination.

The Ministry of Power and Electrification of the Ukrainian SSR was also taken to task for bad planning that had resulted in "extraordinary occurrences" like the schedule to complete a railway network from Vyhoda to the construction site only in 1985, when it was desperately needed immediately, and the lack of such a connection was causing great delays in the supply of materials. One consequence of the lack of a railway was that it was proving impossible to build simultaneously the nuclear power and heating plant, and residential buildings for those who were building the plant. All the available transport (i.e., trucks) was taken up with the delivery of sand, broken stone and cement to the plant site, so that there was no time to deliver brick from Odessa to Teplodar to begin building residences.

The lack of a railway also affected the construction itself, entailing numerous and expensive additional shipments of cargo. The parts, which were arriving, could not be made up beforehand at factories, with the result that a large proportion of the expensive equipment was assembled in the open-air at the building site, so that "safety problems" were not difficult to foresee. The trade unions were said to be lacking facilities and prevented from having any real influence over the process, while the workers themselves were still waiting for a place to live. Over 1,000 people were "standing in-line for residences" but the preparation of the ground on which the workers' "movable cottages" were to be sited had not taken place for almost a year. "This situation," said the author, "cannot continue for long."

Unfortunately for those constructing the Odessa station, the situation does not seem to have improved with time. A report of January 1986 cited Dubensky as saying that the optimism present at the laying of the first metre of concrete in 1981 had evaporated, and the schedule had not been fulfilled. In the first place, he continued, those involved in the construction had begun without a definite technical plan. Sluggishness had revolved around many questions of planning and scientific research. The problems were due, he said, partly to the organization of labour and partly to "material-technical weakness." According to the brigadier of the construction brigade, A.I. Dzhur:

We are being hindered by incorrect planning: people only live for today. But tomorrow the Flood will come. Our leaders must use techniques and plan their work more tightly.[63]

In short, the Odessa nuclear power and heating plant is facing serious difficulties, and the plant, by all accounts, is being built in a haphazard fashion that even according to a Soviet account may be a threat to safety.

Khmelnytsky Nuclear Power Plant

In addition to the South Ukraine nuclear power plant, the Khmelnytsky station in Western Ukraine is also being built with East European help for East European needs. Construction began in 1978, and the station, near the new town of Netishyn, was scheduled to come into operation in the Eleventh Five-Year Plan, but has fallen well behind its programme for completion. Ultimately, it is to have a capacity of 4,000 megawatts, based on four VVER 1000 reactors, and at its peak should produce about 6 billion kilowatt hours of electricity per year.[64]

According to a Polish account, the USSR is the biggest investor in the station, and Czechoslovakia, Poland and Hungary have also made contributions. The Soviets are also responsible for the planning of the station, and are supplying the uranium, while other countries are supplying a variety of machines. After the USSR, Poland appears to have the largest stake. The Polish "Energopol" firm has a total of 2,700 specialists at the plant site drawn from various parts of the country, who are involved in the building of pipelines and canals to supply the reactors with water, in addition to the construction of housing and recreational buildings.[65]

Khmelnytsky is one of thirty Ukrainian enterprises in which Polish workers are participating.[66] Reportedly one of the stimuli for shock work is the "friendly international socialist competition" that is taking place between the Poles and Soviet construction workers at the site. This has proved to be a "solid foundation" for the most rapid construction of all objects.[67] A power supply line, 396 kilometres in length, is being put up between the Khmelnytsky station and the city of Rzeszow in Poland, 114 kilometres from the Ukrainian border. When the line is active, the USSR will reportedly guarantee a 20-year supply of electricity, of which Poland will receive 6 billion kilowatt hours per annum; Czechoslovakia 3.6 and Hungary 2.4 (proportional to the amount of investment). It was envisaged that by 1986, some of the electricity being transmitted on the Khmelnytsky-Rzeszow line would be diverted onto a 110 kilovolt line to serve the southeastern part of Poland. Not only was the station said to alleviate some of the problems associated with nuclear waste disposal in Poland (!), but the report acknowledged that Poland was unlikely to get its domestic nuclear power plant at Zarnowiec into operation before 1990.[68]

Like other stations, the plant was built before suitable accommodation was available for the construction workers. But in the case of the Khmelnytsky station, it led to a managerial crisis. On 19 August 1983, in an article entitled "But the 'Home Front' is Lagging" ("*A 'tyly' ostaiut'*"), *Pravda* noted that although "slightly over 8,000" people worked on the construction, housing was available only for 2,000. Moreover, in one hostel, there was neither heating nor sewerage, while another lacked showerbaths. The Slavutsky urban committee of Khmelnytsky oblast was rebuked for a total lack of attention to the interests of the nuclear plant workers and for falling so far behind schedule in the construction of residential buildings.

Two months later, on 26 October 1983, *Pravda* reported that a meeting of the Slavutsky urban party committee had discussed the questions raised by the initial article. First Party Secretary of Khmelnytsky oblast, T. Lisovsky, and the Deputy Minister of Power and Electrification of the USSR, G. Veretennikov were in attendance. The meeting noted that on the day the original article had appeared, about 500 specialists had been diverted from the nuclear plant to work elsewhere in the oblast. A "struggle with embezzlement and mismanagement" was said to have taken an unsatisfactory course, and there were a "whole series of violations" at the station concerning the building of residential, medical and recreational objects. Officials, it was noted in a report of the oblast procurator, had adopted an irresponsible and negligent attitude toward their duties, especially concerning book-keeping and the maintenance of materials.

The outcome was a minor purge of plant officials. Criminal proceedings were instituted against a deputy chief of the plant's construction department, A. Lopachev, and the section chief of the South Energy Construction Mechanization, A. Kushnir, for embezzling construction materials; against the chief of a section of South Energy Construction Isolation, M. Kurgan, for stealing funds; and against the chief of an assembly department, V. Odintsov, and a brigadier, V. Petrushenko, for embezzling rolled sheet metal.

Other officials escaped with severe warnings: A. Lapko, the secretary of the party committee supervising construction work, and the deputy directors of the plant's construction department, A. Dmytruk and A. Shcherbany. The director of construction, A. Hrotsenko, and a deputy construction chief, E. Vazhenov, had to report to the republican and oblast committees of people's control. The list made it clear that most of those involved in building the Khmelnytsky plant were either careless in their duties or had been stealing materials. The consequence was that residential and recreational needs of the workers were neglected. The resultant overhaul of construction officials may have had some part in the long delay in bringing the Khmelnytsky station on-stream. It may also, however, have been part of a general campaign against embezzlement of public property that was taking place in the USSR under the new General Secretary, Iurii Andropov, in which case, the situation at Khmelnytsky may not have been quite as deplorable as painted in the pages of *Pravda*.

The first power block at Khmelnytsky was scheduled for operation in 1986, but in contrast to some other Ukrainian stations also bringing reactors on-stream in this year, there has been no word from either Soviet or Polish authorities to date suggesting that the plan will be met. Under these circumstances, it is not difficult to predict further delays at a station where work has already been so protracted.

Crimea Nuclear Power Plant

The ultimate size of the Crimean station has not been announced in Soviet materials, but it can be estimated at 2,000 megawatts, based on two VVER 1000 reactors. This seems logical from Soviet statements about the plans for completion, which suggest that the plant will reach its full capacity by 1990, following construction work on the *first* reactor in 1986.[69] In other words, the first reactor will not be ready until at least 1987, and it is very unlikely that a plant of larger than 2000 megawatts capacity is envisaged.

Originally, the first reactor was scheduled to come on-line during Eleventh Five-Year Plan. The Soviet media, however, and particularly *Radio Kiev*, have revealed that here, as elsewhere in Ukraine, a familiar tale of work defects is to be told. "Regrettably," announced *Radio Kiev* in February 1986, "the building work is unsatisfactory." The necessary "acceleration work" had not been procured. The collective at the site had not touched over 200 million rubles designated for the construction, which, as the radio station pointed out, was a huge sum of money. Although the 1986 year was now well under way, the plan for 1985 had still not been completed. According to chief engineer, I.F. Shpak:

> I would like to turn attention to the low fulfillment of discipline at all levels. The time period for the fulfillment of work often drags on, and frequently the decisions accepted remain on paper. For example, the time scheduled for the purification of the construction has been extended many times.[70]

The following month, the radio station returned to the attack and announced that the state of construction work at the Crimean plant "remained unsatisfactory." The building plan for the first three months of 1986 had not even been fulfilled by 50 per cent, principally because of a problem with "toiling resources" and the "very considerable fluidity of cadres."[71] In plain language, this indicated that workers were unwilling to stay very long at their posts, and the implication is that working morale here is at a low level. Thus the Crimea can be added to the list of Ukrainian nuclear plants in trouble at the time of the Chernobyl disaster. Soviet reports revealed once again a chapter of problems that as yet have shown few signs of successful resolution.

Kharkiv Nuclear Power and Heating Plant

The organization responsible for constructing the Crimean plant is the Kharkiv-based Institute for Nuclear Thermal Electricity Planning (*Atomteploelektroproekt*), which also drew up the plans for the South Ukraine, Zaporizhzhia, Bulgarian and East German nuclear plants. In January 1985, it was announced that this institute had a new task: a nuclear power and heating plant that was to be constructed in the city of Kharkiv itself, with a total capacity of 2000 megawatts.[72] At the end of 1985, the Soviet news agency *TASS* announced that actual building work was to begin in 1986.[73]

According to a Soviet Ukrainian newspaper, the number of construction workers at the site rose by January 1986 from 230 to 600, bolstered partly by engineers and specialists from the Zaporizhzhia station, who are supervising the work. Ten organizations are working on the project, including 500 workers from the Kharkiv nuclear planning institute cited above. By means of what is described as a "technique of parallel planning," the planners intend to cut about three years off the normal construction time. The location may play a role in this since the turbines for this plant, as for many others throughout the USSR, are to be provided by the "Turboatom" factory in Kharkiv.[74]

In contrast to most other Ukrainian nuclear facilities, residences at Kharkiv are evidently to be finished before the main construction work gets underway. The dictum being followed here is "An atomic city begins with its people," although this sort of wisdom has not always predominated hitherto in the thinking of Soviet planners. Within five years, 150,000 square metres of living space is scheduled.[75] As for completion, the plant is supposed to be in full operation by 1995. At present, there is said to be enough heat and electricity to supply Kharkiv, a city of 1.5 million inhabitants that is the second largest in the Ukrainian SSR. But the requirements are constantly increasing because of both industrial and population growth.[76] In contrast to the Odessa station, where the nuclear heat and power plant is to replace the old boiler-houses that evidently pollute the atmosphere of the south coast city, the Kharkiv plant seems designated to play a supplementary role, and mainly for industrial purposes.

As with Odessa, the location of any sort of nuclear plant so close to a major city—and all the nuclear heating plants are being built within 20 kilometres of the various city centres—would seem to pose some kind of environmental or safety hazard. Soviet reports, however, have dismissed such questions out of hand. According to the Ukrainian newspaper *Radianska Ukraina*, research has demonstrated that the level of radiation in the region around the station will be "considerably less" than the rays of the sun. Moreover, the work locale is apparently considered so safe that the workers there do not receive bonuses of "danger money" because "the work is not considered dangerous."[77] Whether the population in the vicinity would concur with this assessment may be another matter.

Cherkasy Nuclear Power Plant

In June 1985, the future construction of yet another nuclear power plant in Ukraine—near Chyhyryn in Cherkasy oblast—was announced in the Soviet press, also using the Zaporizhzhia-style "unified flowline system."[78] According to a Western source, the site of the plant was formerly assigned for the building of an oil-fired central electric station during the 1975–80 Five-Year Plan, but subsequently this project was abandoned. The same source also informs that the plant is to be located on the south side of the Kremenchuk Reservoir on the Dnieper River, and that it will most probably consist of water-pressurized reactors with an ultimate capacity of 4000 to 6000 megawatts.[79] The plant is Ukraine's ninth announced nuclear facility.

This survey of Ukrainian nuclear power plants has indicated that while Ukraine's nuclear stations are being built at the most rapid rates possible, and at more and more locations, problems have multiplied. The problems that have been illustrated here raise a more general question that needs to be answered before turning specifically to the giant nuclear power plant at Chernobyl: are Soviet nuclear power plants intrinsically unsafe? Do the shortcuts taken to accelerate the completion of reactor sets pose serious danger inherently not only for the peoples of the Soviet Union, but for the world at large?

TABLE 1   CAPACITY OF NUCLEAR POWER PLANTS
IN UKRAINIAN SSR, JUNE 1986

Location and Year	Capacity in Megawatts	
	Proposed	Existing
Chernobyl (1977)	6,000	3,000
Rovno (1979)	2,880	880
South Ukraine (1982)	6,200	2,000
Zaporizhzhia (1984)	7,000	2,000
Khmelnytsky	4,000	—
Crimea	2,000	—
Odessa	2,000	—
Kharkiv	2,000	—
Cherkasky	4,000–6,000	—
Kiev	at planning stage	

SOURCES: *Pravda*, 26 July 1985; *Izvestiia*, 20 June and 5 July 1985; *Robitnycha hazeta*, 26 July and 22 October 1985; *Radianska Ukraina*, 4 January 1986; *TASS*, 1 January 1985; *Soviet Geography* (May 1983): 394, 396 and (October 1985): 646.

CHAPTER FIVE

Safety in the Soviet Nuclear Power Industry

The Chernobyl accident has raised many questions about the safety levels at Soviet nuclear power plants and related installations. In view of these queries and the obvious problems at Ukrainian nuclear stations in particular, it is necessary to discuss briefly the general situation throughout the Soviet Union. Do Soviet nuclear power plants constitute an ecological hazard, particularly in the European part of the country? Was Chernobyl a result or symptom of a malaise in the industry as a whole, in which workers at nuclear plants are facing everyday dangers, or was it rather a unique chance occurrence, which, logically, should never have occurred, given the safety mechanisms already in place?

The Soviet attitude toward this question has varied from apparent concern and preoccupation to one of virtual disdain that such a problem even exists. Quite often the approach depends on the audience: the former predominates in domestic discussions or in newspaper articles that are intended primarily for a "home" reader, while the latter is usually to be found on *Radio Moscow*'s World Service and in *TASS* statements. At the same time, it appears that the casual (or disdainful) approach to the question, before the disaster at Chernobyl, had gained the upper hand. Since the Soviets have discounted the Urals disaster and have never acknowledged minor accidents in their nuclear industry, they have been able to claim a 100 per cent safety record. And in truth, the Urals disaster of 1957–8, while it may have been of catastrophic proportions, is of little relevance to safety standards in the 1980s, unless one adheres strongly to the view that little has changed in Soviet society over a thirty-year period.

To turn to a few examples. The authors of the informative work *Energetika SSSR* (1981), A.M. Nekrasov and A.A. Troitsky, informed readers that control over radiation safety was being monitored at every

95

atomic energy station by a special dosimetric service, which encompasses all the vicinity of the station and the surrounding area within a radius of 30–40 kilometres. They noted that the service was observing the atmosphere, water and river pollution, but that over the course of "137 reactor years" worked by Soviet nuclear plants by 1 January 1981, the concentration of radioactivity in the region of nuclear plants was always below the permissible norms and hardly ever differed from the normal background level (what this level was it did not specify).[1]

Two years later, *Novosti* affirmed that "present-day Soviet nuclear stations are the safest ecologically." The amount of radiation taken in by personnel working at the site, it continued, does not exceed one-hundredth of the allowable dosage established by doctors, and moreover, one received a higher dosage sitting before the television set than from a Soviet nuclear power station. Scientists, the news agency pointed out, have estimated the probability of an accident at a nuclear plant involving a radioactive discharge at one in one million, i.e., the worker has as much chance of being struck by lightning.[2] This report, which was published in English, typifies the casual approach to safety, but it would be erroneous to accept it at face value, as implying that the Soviet authorities have minimal interest in safety questions.

According to *Radio Moscow* spokesperson, Boris Belitsky, who hosts a regular Science and Engineering programme, the safeguards imposed at all Soviet nuclear plants take up about 50 per cent of the overall cost of the station. "There is no intention," he stated in September 1983, "of economizing at the expense of safety." In his view, it was essential to ensure that personnel at the sites received adequate training, and that departmental interests did not take priority over those of the public, particularly in light of the rapid expansion foreseen for the industry.[3] Similarly, after the American journalist Jack Anderson had written an article in the *Washington Post* accusing the USSR of handling their nuclear industrial waste in a "criminally negligent manner," an irate article by V. Mikheiev appeared in *Izvestiia* refuting Anderson's assertions and commenting that in the entire period of the history of nuclear power in the Soviet Union, no measures have ever been needed to protect the population as a result of an accident or from radioactive products leaking out beyond the vicinity of a station.[4]

According to a prominent Soviet official, control over nuclear plant safety is divided between three organizations: the State Committee on Supervision of Safe Operations in Industry and Mining (USSR Council of Ministers), which is concerned with rules and engineering safety standards in design, building and performance; the State Nuclear Safety Inspection, which supervises nuclear safety in the above areas; and the

State Sanitary Inspection of the USSR (USSR Ministry of Public Health), which is concerned with radiation safety. Each of the three organizations is reportedly guided by a set of regulations, all introduced in the 1970s. The most important of these is the *General regulations to ensure the safety of nuclear power plants in design, construction and operation* (1973), which encompasses all the commercial reactors currently in operation in the USSR: graphite-moderated; water-pressurized; fast breeders; and district-heating reactors.[5]

In October 1984, the USSR State Nuclear Safety Inspection committee carried out what was described as a "wide-ranging survey" which reportedly confirmed that Soviet nuclear power plants are safe and reliable, according to the USSR Deputy Minister of Power and Electrification, G. Veretennikov. As far as nuclear waste was concerned, the system adopted by the USSR was said to be "reliable."[6] Veretennikov did not say in this instance what this system was, nor how the survey was conducted.

In February 1985, the USSR signed an agreement with the International Atomic Energy Agency, the UN body based in Vienna, placing some Soviet civilian nuclear installations under international inspection. A. Petrosiants, the Chairman of the State Committee of the USSR for the Use of Nuclear Energy, declared that the agreement had been reached in the interests of non-proliferation,[7] but at the same time, it also served the useful purpose of rebutting Western accusations about the secrecy of Soviet nuclear installations. Further, it implied that the Soviet authorities had nothing to hide as far as safety standards were concerned, since the IAEA inspectors could hardly fail to have noticed defects had they been present. In the event, however, the three inspectors visited only the large Novovoronezh nuclear power plant, a model specimen, and a small research facility near Moscow,[8] so the visit—in terms of a check on Soviet facilities—was more of an occasion for raising the prestige of the Soviet authorities than a major outside scrutiny of the nuclear industry in the USSR.

Occasionally, Soviet scientists and others have not always accepted the official line on safety levels, although the protests have never been of long duration. The question of waste, for example, has remained a problem, and in the 1970s a variety of possible solutions were being tried out in the CMEA countries (clearly at the Soviets' behest) in the earth's upper strata and in salt formations. According to a Western writer, the Soviets themselves have experimented with the disposal of liquid nuclear waste in deep water-bearing seams underground. He points out also that solid wastes are encased in bitumen and kept in concrete containers above or below the ground until the radioactivity ceases.[9] In *Pravda* in

1981, the noted academician Petr Kapitsa even went so far as to suggest rocketing the waste into outer space, but admitted that this method was not yet foolproof.[10]

In 1983, a fairly frank interview with the First Deputy Minister of Power and Electrification of the USSR, G. Shasharin, appeared in the Soviet press. In response to the question: "What influence does an atomic energy station have on the environment?," Shasharin replied:

> In carrying out the nuclear fuel cycle, man first extracts the uranium ore, refines it into nuclear fuel, removes the depleted fuel, and buries the radioactive waste. In doing all this, it is clear that man is interfering with nature and disturbing the natural balance by contaminating the environment....Atomic energy stations are not totally harmless. Virtually any human activity has some degree of danger and affects nature, but the only potential danger of atomic energy stations is the possibility of a leak of radioactive coolant and other elements into the environment.[11]

Having made these admissions, however, which are somewhat unusual for Soviet ministers, Shasharin maintained that the same imposition on the environment occurred during the non-nuclear fuel cycle and that the main types of coal contain harmful substances that are discharged into the atmosphere along with stack gases. He also stated that the level of contamination of the environment from radioactive substances had declined over the past 15–18 years because of a decrease in nuclear weapons testing. In conclusion, Shasharin stated that "quite a few problems exist" in the nuclear industry, including ecological ones, and especially questions concerning the storage, transport and processing of spent fuel and the disposal of radioactive waste.[12]

The problem of storing nuclear waste is by no means unique to the Soviet Union. Yet the Soviet authorities show a surprising degree of difference over whether they have effectively "resolved" this problem. In 1985, Belitsky noted that the Soviets are dealing with the waste by burying it, after proper treatment, deep underground free from any contact with ground water and that safety is monitored by underground instruments that have rendered the underground storage system quite safe.[13] This appears to contradict Shasharin's statement, which admittedly was made two years earlier. In March 1986, H. Marchuk, a Deputy Chairman of the Council of Ministers of the USSR and Chairman of the USSR State Committee for Science and Technology, also stated that the problem of the safe burial of radioactive waste had been solved.[14]

Another official was less certain, declaring that "the results of all scientific and field studies do not yet provide a final answer on the most suitable types of rocks for waste disposal.'' He noted that vitrification is thought to be the best method of dealing with high-level wastes, whereas medium-level wastes in the USSR are maintained in stainless steel tanks, although vitrification is again being considered. As for low-level wastes:

> A universally applicable way of purifying low-level liquid wastes has been developed using a two-stage ion-exchange process. The ion-exchange resins are regenerated and repeatedly used, and the solutions are evaporated. After hardening, the residues are sent for storage, while the water can be used for technical purposes. The final volume of wastes to be stored is only 0.2% of the initial one.[15]

In August 1985, a report surfaced in Sweden from an Estonian defector that a Soviet nuclear waste depository in Estonia had killed at least one person through a leakage of radiation. According to the defector, the waste is stored ''under very primitive conditions'' at the site, which is some 15 kilometres south of Estonia's capital city, Tallinn. A plain concrete bunker is said to be staffed with inexperienced workers who lack dosimetric instruments and are thus unable to monitor radiation levels. The waste was said to come from the Paldiski nuclear submarine base, to the west of Tallinn.[16] The report, from Swedish radio, has not been corroborated by other sources. If it is true—and it seems plausible—then one must have some reservations about whether the waste problem has been ''solved'' in the USSR. In any event, even Soviet writers acknowledge that it will become an increasing problem over the forthcoming decades as nuclear energy continues to expand.

Soviet citizens, while they have never mounted a major protest against the development of nuclear power, in contrast to their counterparts in the West, have at times expressed alarm, and occasionally manifestations of their fears have slipped almost unnoticed into the Soviet press. For example, writing in 1974, V.T. Kizima, the head of the construction unit at the Chernobyl station declared that ''People are familiar with the traditional form of power plant and in time, they will become accustomed to, *and lose their fear of* the atomic plant.''[17] While one might say that public apprehension was only to be expected at a time when Ukraine had never experienced an operational nuclear power plant on its territory, the same can be said of many areas in the Soviet Union today that are facing the prospect of an erection of a nuclear facility in their area for the first time (the city of Kiev, for example).

Similarly, in the interview described above, V. Shasharin was also asked to respond to the following statement:

> The editorial office [of *Sotsialisticheskaia industriia*] often gets letters from people in Odessa and Minsk who are afraid that the nuclear power and heating plants that are being built in those cities will have a detrimental effect on the environment and population.[18]

A discussion about the installation of nuclear heating plants was also held with A. Ie. Shenydlin, the Director of the High Temperature Institute at the USSR Academy of Sciences, in 1984. The interviewer stated that "such construction work would cause alarm among the population," to which Shenydlin responded that "Much of this is the result of emotions."[19]

Recently therefore, there have been indications that some Soviet citizens are afraid of nuclear power, and particularly about the new nuclear power and heating plants that are being established very close to urban centres, in some cases, such as that at Gorky, only two kilometres from the city boundary.[20]

In 1979, Academician N. Dollezhal and doctor of economic sciences, Iu. Koriakin, published a major article in *Kommunist*, the theoretical and political journal of the CC CPSU, entitled "Nuclear electricity: achievements and problems." The article expressed in print in a very influential medium some of the main concerns about the expansion of nuclear energy in the USSR and its effect on ecology, particularly in the European part of the country. While the contents of the article were not original, this was the first time they had been aired in a major as opposed to a strictly scientific journal, and as such, they merit analysis.

The authors noted that because of problems with traditional sources of energy, such as oil and gas, it would be necessary to construct more and more nuclear power plants in the USSR, and that in the Tenth Five-Year Plan, over one-third of the energy growth in the European part of the USSR was to come from nuclear power, which itself was to see a trebling of output. The prospect, which might have occasioned pride in some Soviet circles, did not, however, please Dollezhal and Koriakin. They noted that all the nuclear power plant construction was to take place west of the Volga and the Volga-Baltiskii canal line, where 60 per cent of the population of the USSR resides. Soon, they maintained, this would lead to the exhaustion of the ecological content of the region, measured in terms of the permissible influence on the surroundings of a power plant. More territory would be needed in which to build water reservoirs and living quarters. They used the example that to construct a cooling pool for a typical atomic energy station with a capacity of 4,000 megawatts, an area of 20–25 square kilometres is needed and consequently there is less land available for food production.[21]

Further, the authors perceived a serious water resources problem as a result of the current Soviet nuclear energy programme. Thermal and nuclear energy plants were using up over 100 cubic kilometres of water each year, and every year two cubic kilometres of water were lost through evaporation in the European part of the USSR. The authors postulated that as a result of the industry's expansion, these water losses would double by the year 2000. They also condemned outright the plans to divert the flow of rivers from north to south in the European USSR in order to stabilize the levels of the Caspian and Azov Seas. They considered that the biggest question for the future therefore was the choice of new sites for nuclear power plants, especially for sites "south of Moscow" where there was the biggest demand for energy.[22]

Another problem was radioactivity, not only inside the station, but in the transport system and the railways. The authors pointed out that while nuclear fuel can only be transported in small amounts and the chances of an accident are slight, the problem was likely to increase because of the anticipated growth in the industry. The current method of siting the stations was said to have "had its day." An average of 1.5–2 years was required before the actual work on the plant itself could begin, and there was an outlay of capital expenditure on secondary objects that were not directly related to production.[23]

To all these problems, Dollezhal and Koriakin proposed one radical but not unrealistic solution: to site future stations in huge nuclear complexes in remote areas. They envisaged complexes that would consist not only of groups of nuclear power plants of enormous capacity, but also factories and the means to recycle, store and transport the spent fuel. This sort of concentration would lead to a unification of all the parts of the technological process and a more economical return on investment. More important, it would benefit the ecology of the country. The concept of energy complexes, they concluded, which was also being studied in the United States, seemed to be the best solution for the organization of the nuclear power industry.[24]

The article revealed that in 1979, at least, there was no clear consensus among Soviet scientists as to how best to develop the nuclear programme. From the perspective of hindsight, it appears to be an unexpected attack on the current thinking on the issue. In reality, however, it shows rather that the path that the Soviets have taken in the 1980s was decided upon relatively recently, and after some debate. For our purposes, the significance of Dollezhal and Koriakin's work is its depiction of the ecological damage to the environment brought about by the expansion of nuclear energy in the European USSR. This would apply particularly to Ukraine, where at the South Ukraine and other plants, local water resources and systems are being tampered with, which could have a drastic effect on the surrounding countryside and farmland. (The article's main theme was later repudiated in the Soviet press without mentioning the authors by name.)

Ecological concerns have frequently arisen in other Soviet reports. For example, an all-Union conference was held on the question of the impact of nuclear power plants on the environment as long ago as 1981. Then, as now, the main question was the impact of an increasing number of nuclear establishments in heavily populated areas where water and land resources are somewhat limited,[25] i.e., the same concerns that were expressed in the *Kommunist* article of 1979 discussed above.

The question was raised also in January 1986 by the head of the recently established Scientific Centre for Ecological Problems of Nuclear Energy that has been established at the Academy of Sciences of the Ukrainian SSR, V. Chumak. Chumak noted that while nuclear power plants are cleaner energy sources than stations in which traditional organic fuels are burnt, there is a dual ecological problem of the changes to the surrounding environment and the usage of natural resources that is entailed in the construction and operation of nuclear plants. These stations, he continued, "infringe" on nature and agriculture by consuming huge amounts of water. As a result, before any such plant is brought into operation, several years are expended in which every effort is made to limit the impact on the environment.[26]

The new Centre was given the task of drawing up a prognosis on the question in Ukraine up to the year 2000. It is related also to the development of subsidiary industries within the sphere of the nuclear power plant. Thus at Chernobyl and South Ukraine stations, large-scale fish breeding is carried out in the plant reservoir, and at the latter station, hothouses have also been constructed. There is no question, however, that these small-scale industries, in the long term, can hardly compensate for the changes made to the environment, particularly when the nuclear stations in question are planned to reach huge dimensions.

The impact of nuclear power on the Soviet population would have risen eventually without the impact of Chernobyl because of the construction of nuclear power and heating plants at major Soviet cities. Two plants—at Gorky and Voronezh—are to be heat supply stations (ASTs to use the Soviet initials), while there are plans to build at least six (and probably many more) nuclear power and heating plants (ATETs) at Odessa, Kuibyshev, Leningrad, Kharkiv, Volgograd and Kiev. Generally, these stations are to be located about 30–40 kilometres outside the cities,[27] but Odessa and Gorky stations, at least, are being built much closer than this, as noted above. Ironically, in view of the preoccupation with environmental safety, one of the official raisons d'etre of these stations is to promote "environmental protection."[28]

According to the Soviet news agency, *TASS*, the unblemished safety record of the Soviet nuclear power industry "will make it possible" to start introducing nuclear-generated heating into Soviet cities on a wide scale.[29] Shasharin, the USSR First Deputy Minister of Power and Electrification, pointed out that these new constructions would also be perfectly safe because they produce thermal power in the form of steam and hot water, neither of which can be radioactive, and this "pure water" is channelled so that it never comes into contact with the contaminated reactor water.[30] The statement is to some extent contradicted by the *TASS* assertion above, which suggests that the nuclear heating plants are only being introduced *because* the safety record has enabled it, which suggests that some kind of risk is involved.

The main reason for constructing the new and experimental types of nuclear plants is the saving on fuel that is entailed. The experiment was begun at Bilibino in Chukotka, where a small nuclear heat supply station based on a fast-breeder reactor provides all the local mining enterprises with electricity and services the arctic Bilibino settlement. In this way, the combustible materials traditionally used for this purpose can be diverted to industries in need of them. Tiny Bilibino, however, is a far cry from Odessa or Kharkiv, and Soviet sources acknowledge that their prime concern before establishing such stations is the maximum radiation safety for the nearby residents of large cities.[31]

In a May 1986 Western newspaper article critical of Soviet decisions concerning the locations of nuclear power plants, exiled Soviet geneticist Zhores Medvedev commented that in 1981, at the outset of the Eleventh Five-Year Plan, the Soviet authorities made the "wrong and potentially dangerous decision" to use nuclear plants to heat major cities, and said that no debate had been permitted on the issue.[32] It is too early to make judgments about the safety of nuclear supply or nuclear power and heating stations, although the problems that have arisen in the construction of the Odessa station certainly give cause for concern. What one can say is that the decision to carry out such constructions was based on a growing confidence on the utmost safety of nuclear energy installations in the USSR that can hardly have survived the Chernobyl disaster.

One of the earliest indications that the Soviet nuclear power industry was facing severe safety-related problems was the "Atommash episode" of July 1983. In 1975, the Twenty-Fifth Congress of the CPSU approved the construction of a huge reactor manufacturing complex, Atommash (atomic machinery), at Volgodonsk. It was to produce VVER reactors of 1000 megawatts capacity at a rapidly expanding rate, starting with one in 1978 and rising to a regular schedule of four and ultimately eight reactors a year by 1990. Over the course of the next eight years, the Soviet press carried frequent reports about the slow progress at Atommash. In fact, no reactors were produced before 1980, and subsequently, it has proved impossible for the workers at the plant to produce more than two a year. These problems are connected with the stagnation of the Soviet steel industry which occurred in the late 1970s, but there is also evidence that the building plan was badly thought out and that the work was of such a low quality that, according to one Western source, "some Western experts believe that the reactors are so badly designed that they could be dangerous."[33]

Matters came to a head on 19 July 1983, with the official announcement by the Soviet authorities of the establishment of an all-Union State Committee for the Supervision of the Safe Conduct of Work in the Nuclear Power Industry.[34] While little information was given about the reasons for establishing such a committee on this date, *Pravda* revealed on the following day that a Council of Ministers' decree had been issued concerning the "flagrant violations of state discipline in the draft planning, construction and operation of projects in the city of Volgodonsk." Because of the "unsafe technical decisions in planning" and a lack of control over the quality of the construction here [i.e., at the Atommash plant], G.N. Fomin had been relieved of his duties as head of the State Committee for Construction and Architecture and as First Deputy Chief of the USSR State Construction Committee.

A meeting of the city party officials had taken place in Volgodonsk on 19 July, attended by Politburo Candidate Member and CC CPSU Secretary, V.I. Dolgikh, who made a speech elaborating on some of the "defects" that had occurred in the construction. Having outlined some of the "toiling successes," Dolgikh turned to the shortcomings that were responsible for his visit to Volgodonsk, such as the "substantial defects" in the work at Atommash and the failure to use available resources. He criticized the plant's inability to keep up with the production schedule and alluded to the need to accelerate the training of skilled workers and specialists. Housing and social facilities were also said to be lagging behind schedule.

More serious, however, were questions about the quality of construction work. For a number of years, Dolgikh declared, the Volgodonsk Energy Construction Trust had been guilty of "regular technological violations," had breached the rules regulating building work, had deviated from the design plans and quite often had handed over installations "with major imperfections." "Crude violations of state discipline" had evidently been committed by leaders of certain ministries and subordinate organizations, and the Politburo of the CC CPSU had handed down severe punishments for these transgressions.[35]

In what appeared to be related events on 20 July, the Chairman of the USSR State Committee for Construction, I.Novikov, was sent into retirement "at his own request,"[36] while two major newspapers focused on problems in nuclear power: *Sotsialisticheskaia industriia* printed criticisms of outdated machinery and equipment at Soviet nuclear plants, and *Sovetskaia Rossiia* published requests for better safety gear for those involved in the construction of the nuclear power plant at Balakovo.

These events were not necessarily related to Atommash, and the retirement of Novikov, since he was 77 at the time, may not have been related to anything other than old age. Western analysts speculated at the time that Novikov was removed because of his links with the late Brezhnev—the two were born in the same town in the same year, Dneprodzerzhinsk and had graduated from the same metallurgical institute, for example.[37] But, to use a Stalin phrase, "it was no accident" that it occurred on 20 July after Dolgikh's trip to Volgodonsk.

The Soviet statements set off speculation about what had happened at Volgodonsk among Soviet analysts in the West. Deaths and a major accident were postulated, but as with the Chernobyl accident, a lengthy Soviet silence ensued. On 1 August 1983, Evgenii Kulov, a 54-year old physicist and engineer was appointed Chairman of the new committee to supervise safety in nuclear energy, receiving a promotion from his former post as Deputy Minister of Medium Machine-Building, which he had held for only one year.[38] Within two weeks, Kulov could be heard on *Radio Moscow* refuting Western allegations about the Atommash controversy and about accidents at Soviet nuclear power plants in general:

> Such allegations are pure fantasy. The rate of development of nuclear power is accelerating in the USSR, especially in the more densely populated parts of the country. And its uses are growing. An increasing number of people and organizations are involved in the planning of nuclear power plants...thus it is useful to combine supervision of the plant design, the manufacture of equipment, construction and usage under a united state body.[39]

On the day after Kulov's statement, the British newspaper, *The Observer*, put forward a theory about the events at Atommash. It declared that according to a "scientific source" in Moscow, the accident had been caused by a rise in the water tables in the basins of the Don and Volga rivers. Three lakes had been created in the area following the construction of huge dams that brought about the rise in water levels. *The Observer* article alleged that those officials responsible for siting Atommash had apparently ignored a slow rise in the water-table of the region, and that at some point, this water had undermined the plant's foundation, and also threatened other major industrial sites in the area.[40] It seemed therefore that nature had exacted a revenge on the nuclear industry.

The *Observer*'s speculations were evidently well founded. Almost a month later, the Soviet authorities acknowledged that "those who should have known better" had failed to take into consideration the "peculiarities of the subsoil" around the Atommash plant. As a result of a flagrant planning and construction error, certain buildings fell victim to subsidence. The rumour went around, said the report, that "Volgodonsk is adrift." Evidently a number of workers left the plant, while others wrote letters to friends advising them not to come to the plant because "no one knows what will happen." Following a discussion in the CC CPSU Politburo, workers were said to be labouring furiously to rectify the situation.[41] The above account, it must be added, did not tell the full story, which to date has never been revealed. Nevertheless, it demonstrated the chronic problems of planning that had brought about such a fiasco.

Although Atommash is now believed to be functioning again, it has never lived up to its expectations and the prospect of an output of eight reactors a year still seems well beyond its scope. The nature of the accident at a nuclear installation, which fortunately does not use radioactive materials, raises the question of whether the Soviet nuclear industry generally has been subject to the sort of planning, construction and supply disorders that have so plagued the Ukrainian sphere and which were clearly revealed at Atommash.

To begin, one can cite two Western opinions about the way in which the Soviet nuclear industry operates: the first from a journalist quoting "Western experts" and the second from a scholar at London's Uranium Institute. According to a report in *The New York Times* of 1 May 1986, Western nuclear specialists have stated that the USSR "has the worst safety record of any nation" including the rest of the "Soviet bloc" and developing countries. Among the alleged defects of the Soviet industry, the following were listed: a "hazardous" reactor design; a reduction on safety standards for the sake of economy; unsatisfactory methods to contain radiation, cool nuclear fuel, utilize computers and provide for secondary safety measures; poor evacuation procedures; the location of reactors in populated zones; and the lack of public scrutiny over safety measures.[42]

Some of the above may have been postulated with the Chernobyl accident in mind, and further, the date of the article, 1 May, precludes a dispassionate approach as it precedes the dates on which the Soviets, finally, were prepared to provide information about what had occurred at Chernobyl. In short, had the article in question appeared on 1 April rather than 1 May, it may not have been quite so critical, although there is little doubt that some of the criticisms would appear to be justified by the evidence available.

In his careful study on "Nuclear Power in the Soviet Bloc," W.P. Geddes of the Uranium Institute notes that the nuclear industry is adversely affected by general problems of Soviet planning, which render it impossible for the planners to "co-ordinate every aspect of each sector of the economy." Bottlenecks result as it becomes more and more difficult to extract the various parts needed from low-priority sectors of the economy.[43] Geddes maintains that the centralized planning system of the USSR is inappropriate for the development of the nuclear industry, in which the components are highly specialized and must be made to particular specifications. As seen earlier, the Soviets have sought a way out of this dilemma by serializing production of nuclear components at Zaporozhzhia and elsewhere.

There is ample evidence to suggest from Soviet accounts that the quality and the methods of construction of nuclear facilities are inadequate. At Balakovo, for example, in May 1982, the chief engineer said to the representative of the firm supplying the pipes for the nuclear power plant:

> We have examined your pipes with ultrasound—complete junk. There are even defects that can be seen with the naked eye. Moreover, the metal is not of the specification called for in the plan. After all, it is a nuclear power plant![44]

Four years later, Moscow television featured the Balakovo nuclear power plant in a programme that combined information with sarcastic criticism:

> How difficult it was for those who built Balakovo atomic energy station's first power unit to make 15,000 supplementary holes in such a monolith, for some reason unforeseen in the original project. And such blunders on the part of the planners, and disruptions in the supply of equipment, have been piling on top of our own mess here. Lack of experience has taken its toll—the collective has not yet built any atomic stations.[45]

The programme went on to assert that "violations of labour and technological discipline had made themselves felt" and that various groups of workers who belong to a single party organization were tackling questions about the building not with each other, but through higher bodies in Moscow—hence a classic example of the dilemmas brought about by the centralized chain of command. "Interderpartmental barriers," it was reported, "tend to be stronger than concrete." At Balakovo two 1000 megawatt units were scheduled to be in service by the end of 1986, but the first had only just been added to the grid, and the programme stated that the timetable "should probably be revised."

The Balakovo case is not untypical. Because of the shoddy workman-

107

ship and engineering at many plants, shock workers (*udarniki*) have been sent from site to site to resolve some of the problems that have arisen. In turn, this has weakened the workcrews elsewhere.[46] At the Smolensk station in 1981, *Radio Moscow* reported that there was a great shortage of skilled personnel and that in order not to stop operations altogether, it had been necessary to employ workers without the proper qualifications.[47] Similarly, a major daily newspaper dwelled on the "litany of complaints" about the reactor construction industry in this same year, and commented that often work crews did not have sufficient training for the difficult job of installing reactors.[48]

Like any new industry, time is required in the nuclear sphere for the development of skilled personnel. In the Soviet case, it seems that the rate of expansion of the industry has exceeded the number of cadres available. There are a number of references in the Soviet press to the shortages of qualified people,[49] and one Western scholar has suggested that the advertisements in the press for workers at nuclear power plants indicates a reluctance on the part of Soviet workers to take up posts at such locations.[50] This is a problem that is evident at every nuclear power location from Chernobyl to Balakovo, and it poses a major safety hazard.

In the summer of 1985, for example, a report from Ukraine dwelled on the shortage of qualified cadres at nuclear power plants. The majority of the skilled workers trained in the republic had attended the Odessa polytechnical institute, which had been founded in 1975. Because of the growth of nuclear power plants, however, the authorities had been obliged to start preparing cadres at the Kiev polytechnic. Nevertheless, a key role in basic construction work fell to unskilled labour, notably students. "This summer," the report noted, "toiling squads of students" are working at the Odessa nuclear power and heating plant, and the atomic energy stations at Rovno, Chernobyl, Khmelnytsky, South Ukraine and Zaporizhzhia.[51]

Added to the skilled labour problem and closely related to it is the defective equipment that often arrives at the Soviet nuclear plants. At the Ignalinsk nuclear plant in Soviet Lithuania, where graphite-moderated reactors of 1500 megawatts are being installed for the first time anywhere in the USSR (or the world), it was reported that the first generating set, which came into operation on schedule on 1 January 1984, had had to be shut down because of construction defects and because the equipment was of poor quality and supplied "in the wrong order." Repair work was said to be of poor quality, while the station's computer system, completion of which had been delayed, was also unsatisfactory. At the second power block, construction was lagging because of the delay in receiving some technical equipment, and some "violations of safety rules and labour discipline" had occurred.[52]

After Chernobyl, the London *Daily Telegraph* referred to Ignalinsk as "a disaster waiting to happen" because of the safety violations cited in Soviet reports.[53] While this may be true, the Vilnius example could be applied to virtually any nuclear power plant in the USSR. These same drawbacks are cited ad nauseam in the Soviet press.[54] A lack of quality control, an unskilled and dissatisfied workforce, supply problems, defective equipment, lagging construction, plans arriving late, design changes and cost overruns sum up the main difficulties facing the Soviet authorities in the nuclear sphere. Such defects, while serious in any industry, are of critical importance in such a unique sphere as the atom. What, then, of the reactors themselves? Is their design manifestly unsafe? And why have the Soviets advanced with two designs rather than one? Is the Soviet RBMK now antiquated?

According to a handbook of the Leningrad nuclear power plant (handed out to visitors), the basic dimensions of the RBMK 1000 reactor, upon which Chernobyl was also modelled, are as follows:

Electric capacity: 1,000 megawatts
Heating capacity: 3,200 megawatts
Process channels: 1,693
Reactor charge: 180 tons of uranium
Initial enrichment: 1.8 per cent
Burn-up, megawatts/day/kg: 18.5
Temperature in degrees celsius: Saturated steam: 284
Reactor inlet water: 270
Weight of graphite brickwork: 1,700 tonnes.
A Soviet manual, cited in the Ukrainian press, noted that the reactor is located in a concrete well 21 metres square and 25 metres high.[55]

After the Chernobyl disaster, several Western newspapers quoted nuclear experts who criticized the RBMK 1000 reactor used for the generating sets there. For example, the London *Daily Telegraph* reported that Dr. Tom Marsham of the Atomic Energy Authority visited the nuclear power plant at Leningrad, which was the first plant in the USSR to use a 1000 megawatt graphite-moderated reactor (1973), and declared that "Those reactors seem to be designed with a potential for every possible sort of fault." One of Marsham's premises was that when graphite rods overheat, it is particularly difficult to cool them down. In addition, like many Western experts, he criticized the absence of a leak-proof containment building around the reactor.[56]

The lack of containment has never been perceived as a major problem by Soviet officials. Indeed, A.M. Petrosiants, the Chairman of the USSR State Committee on Atomic Energy informed a U.S. official in 1976 that there was no need for containment buildings. If customers required them, then they would be built, he noted, but the Soviets themselves would not construct them for domestic use.[57] After the nuclear accident at the Three Mile Island plant in 1979, the Soviets began to put reinforced concrete containment structures around their water-pressurized reactors, but the RBMKs were still built with only a relatively thin protective shell, as was evident at Chernobyl.

According to the *Financial Times*, the RBMK originally was built along the same lines as the U.S. plutonium-producing reactors built during the Second World War.[58] Another Western newspaper has also stated that the Soviet authorities began to build this kind of reactor in the 1950s because they could fulfill a joint function: the production of electricity and plutonium for the Soviet nuclear weapons programme. The neutron spectrum of graphite reactors, it declared, enables the production of high-quality plutonium, and in fact each of the USSR's twenty graphite-moderated reactors "turns out large amounts of plutonium on a continuous basis."[59] In short, the opinion one might gain from the combined information given above is that the Soviets kept to a fairly dangerous design because they wanted to give priority to a weapons programme.

This may or may not have been true of the 1950s. There is little evidence to suggest that today, however, that all Soviet RBMKs are used for weapons' plutonium manufacture. Clearly some have been designated for this purpose, but there is also some evidence that the Soviets have continued to manufacture reactors of this design for other reasons (which will be dealt with below). One other critique of the RBMK that should be mentioned at the outset, however, is what one source has termed its "uniquely bad combination" of moderator and cooling method, i.e., the use of graphite and water respectively. Whereas the U.S. reactors use water as the moderator, graphite is known to burn for long periods. The British, on the other hand, like the Soviets, use graphite as a moderator but carbondioxide rather than water for cooling purposes—the graphite does not burn in carbondioxide.[60]

The Soviet authorities have two main reactors that are produced at two locations: the Atommash plant at Volgodonsk—as noted above—produces VVER 1000 reactors, while the Izhorsky engineering works in Leningrad manufactures the RBMKs, which at present are being brought out in two sizes: 1,000 and 1,500 megawatts. The larger type, which is already in operation at Ignalinsk, is produced on the premise that "the more powerful the reactor, the more economical is its performance."[61] Although the RBMK remains the predominant reactor in terms of capacity, it is the VVER that is being used in the majority of new nuclear power plants being built in the USSR, in addition to Eastern Europe, which uses the VVERs exclusively. The RBMK has to date never been exported by the Soviets, nor has its design been repeated elsewhere. Because of its restriction to domestic use, it has been termed the "Soviet reactor."

Appearing on Soviet Television in May 1986, L.P.Feoktistov, the Deputy Director of the Kurchatov Atomic Energy Institute at the USSR Academy of Sciences, stated that the RBMK reactor was of a pressure-tube type, based on some 1,600 tube pipes (channels) made out of zirconium. The pipes, he noted, contain heat-emitting elements made up of uranium. Within them flows water, which cools the uranium, as a result of which energy is produced through fission. Then the water heats up and turns into steam. In this form, it enters the turbines, which, in turn, generate electricity.[62]

According to another Soviet source, its foundation on channels gives the RBMK several benefits in addition to the possibility of producing both electricity and plutonium:

In addition to a good neutron balance and a flexible fuel cycle, channel reactors make it convenient to monitor the state of fuel elements of individual assemblies and the integrity of the channels; fuel assemblies containing elements with ruptured cladding can be replaced without shutting down the reactor; as a result of increasing the number of loops and decreasing the diameter of the pipelines the dangerous consequences of the rupture of pipes are diminished; the unit power of a reactor can be increased without complicating the emergency cooling system.[63]

A Soviet official also points out two other advantages of the RBMK: its components can be produced at factories already in existence and do not necessitate purpose-built enterprises; and the reactor's make-up and its on-load refuelling system enable a high efficiency useage of low-enriched fuel. "The discharged fuel has a low fissionable material content, the burn-up is high, and the plutonium produced in the fuel is utilized."[64]

The design is perhaps crucial, particularly in view of the Soviet tendency to increase the dimensions of their reactors over a number of years: the VVER thus began with a 210 megawatt capacity, increased to 365 and 440 megawatts, and expanded ultimately to 1,000 megawatts in the 1980s.[65] In turn, the RBMK has been expanded in size from 1,000 to 1,500 megawatts at Ignalinsk and there have been some indications that even larger reactors were in the offing before the Chernobyl accident occurred. As the size increases, the costs of construction reportedly decline correspondingly. Consequently, a containment shell, which has been built over the VVER 1000 reactors from 1981 onward, is financially impractical on the RBMKs.[66] Soviet thinking on this question—which

111

proved to be a crucial factor in the Chernobyl accident—is obtuse. We noted above that one Soviet nuclear official declared that such containment was not necessary. But there have been a number of Soviet reports extolling the benefits of containment shells, such as the one built for the fifth VVER 1000 set at the Novovoronezh nuclear power plant.[67] In short, once they were built, the Soviets either perceived or gave an impression that they perceived their benefits.

A British source points out some other factors that may have prompted the Soviet authorities to continue with the RBMK design. It notes that the RBMK is constructed on the site "with modular sections" based on standardized components that can be brought to the site by rail. The plants based on this design on built in pairs, "with interchangeable auxiliary systems." The total time period for the construction of a pair of reactors is said to be only 7.68 years and declining.[68] The construction of twin reactors may be preferable in terms of the time factor, but their use of the same generating room and close proximity proved to be a major hazard during the Chernobyl affair when there was an immediate danger of the fire spreading from the fourth to the third generating set.

For the immediate future, the Soviets will likely persist with both RBMK and VVER designs, with the latter taking up an increasing share of the total capacity. For the more distant future, however, Soviet reports have emphasized that they are to rely to an increasing extent on the so-called "fast-breeder reactors," an area in which some Soviet reporters have boasted that their technology is ahead of that of the United States.[69] The first large-scale fast breeder—of 350 megawatt capacity—was developed at the Shevchenko nuclear power plant, on the Mangyshlak peninsula of the Caspian Sea, in 1973. The fast-breeder type reactor, which was pioneered at Bilibino, has the advantage of using less uranium and at the same time producing sizeable quantities of plutonium during the nuclear reaction.

In 1985, the official Soviet position on the fast breeder reactor was elaborated by Dr. V. Legasov, Deputy Director of the Institute of Atomic Energy of the USSR Academy of Sciences in Moscow.

The present conception of nuclear reactors necessitates mining a great deal of uranium, since only a small share of it is actually utilized. This is because the naturally mined uranium contains less than 1 per cent of isotope uranium-235, which is used as nuclear fuel. Therefore only a tiny fraction of the uranium mined is actually put to use in conventional reactors. This difficulty can be overcome by fast breeders. Unlike conventional reactors, which use thermal, or slow neutrons, fast breeders use isotope uranium-238, which makes up over 99 per cent of all the uranium in the world.[70]

Legasov pointed out that the reaction that takes place when fast neutrons bombard atomic nuclei to produce energy also produces a new element—plutonium-239—and that a reactor can be designed to produce more of this new element than it burns uranium. Because of this development, the term "breeder" is used as an appellation. In the future, he said, larger breeders would be installed at Beloiarsk nuclear power plant and other locations, first of 600 megawatt capacity and later of 800 and 1,600 megawatts. In addition they would be used to provide heat for cities as well as electricity.[71]

The era of the fast breeder remains in the future, however, and this type of reactor is not expected to be a factor in the Twelfth Five-Year Plan to any significant degree. As far as the safety factor is concerned, one can make two conclusions. First, that while the RBMK may not be inherently unsafe as a design, as some Western analysts have claimed, it has been used less for its safety features than for its economic advantages. On-line refuelling makes containment shells unviable, and such containment, in turn, would serve to restrict the size of the RBMK reactor. The VVER is less compact, and the capital costs involved in its construction are higher. Also the VVER 1000 is built with large pressure vessels that are not easy to transport. According to one source, the 1,000 megawatt vessel is 11 metres high and 4.3 metres in diameter.[72] Thus it is possible that in view of transportation problems, the water-pressurized reactors may be close to their practical capacity at 1,000 megawatts.

Moreover, the problems at Atommash may mean that the Soviet authorities could not move to a more complete dependence on the VVER had they the inclination to do so. For the time being, the RBMKs will persist in the nuclear programme, despite some apparent flaws in safety design.

The subject of accidents at Soviet nuclear power plants and at nuclear reactors generally is rarely mentioned in Soviet reports or by Soviet spokespersons. Aside from the reports of the Urals disaster by Medvedev and others—and the standards of the 1950s are not necessarily applicable to those of the 1980s—little is known about other incidents. Yet some have occurred and should be cited in brief to counter the notion that before Chernobyl, Soviet nuclear power installations had a perfect safety record.

At a meeting with Representative Robert H. Michel of Illinois in 1979, former USSR Minister of Power and Electrification, Petro Neporozhny, reportedly said that there had been "several accidents" at Soviet nuclear power plants, including one involving an explosion and one a leak of radiation. According to Michel, the minister on one occasion donned a radiation suit and personally visited the accident area.[73] There is little reason to doubt the authenticity of this account, which is quite plausible given what has been shown above concerning the immense problems that have occurred and are still taking place in the construction of such facilities.

There have also been a number of second-hand Western reports of incidents at Soviet nuclear facilities. While these are outnumbered by the 20,000 plus "incidents" that the Soviets claim have occurred at U.S. facilities, they are nonetheless worth citing.

According to the U.S.-based Fund for Constitutional Government report, the Soviet icebreaker *Lenin* experienced a "reactor casualty" that led to a substantial leak of radiation in 1966 or 1967, after which the ship was said to have been abandoned for over a year. Also at sea, on 11 April 1970, the same report stated that a November-class submarine experienced a "casualty in the nuclear propulsion system," as a result of which it sank 300 miles off the Spanish coast with 99 crew members aboard.[74]

In 1969, a reactor at the Novovoronezh nuclear power plant had to be shut down for repairs to its shield.[75] In either 1973 or 1974 (depending upon which source is consulted), the cooling system at the fast breeder reactor of the Shevchenko station broke down, and there was allegedly a "severe explosion" there.[76] Czechoslovak dissidents have reported two serious accidents at the Jaslovske Bohunice station in 1976, which is based on Soviet-designed water-pressurized reactors.[77] Finally, in 1981, there were reports in the West about a pressure build-up at the Rovno nuclear power plant that damaged a steam generator and closed the plant down temporarily. At that time Rovno consisted of one 440 megawatt reactor.[78]

To these specific accidents can be added the radiation leaks at the Estonian waste dump described earlier, and the verifiable catastrophe at Atommash. The Chernobyl disaster has in fact added credibility to some of these earlier reports, as has the Soviet insistence on dwelling on incidents at nuclear power plants outside the Soviet Union, which gives the impression that the accuser has something to hide. At the same time, it would be wrong to maintain that in light of the above chronicle of problems and the few noted previous incidents that Chernobyl was inevitable. The events surrounding the accident, and the prehistory of the plant will demonstrate that it was rather a symptom of a general malaise; of problems that have pervaded virtually all Soviet industries, of a low working morale, and of pedestrian if not outmoded safety measures in the event of a worst-case accident. None of the above may have caused the accident, but they are all closely identified with it nevertheless.

CHAPTER SIX

The Chernobyl Disaster

The Background

The twelfth-century town of Chernobyl (Chornobyl in Ukrainian) occupies the junction of the Prypiat and Uzh Rivers, 133 kilometres north of the Ukrainian capital city, Kiev. It stands in an area that has long been considered a problem for agricultural cultivation; of sandy podzolic soil, peat bogs and marshland. And while by Soviet standards, it is located close to major population centres, such as Kiev, Chernihiv and the city of Gomel in Belorussia, it is in one of the most sparsely populated areas of Ukraine. Before the construction of the nuclear power plant, what little industry there was consisted of food processing and small-scale shipbuilding; the main agriculture consists of flax and potato growing, and only about one-third of the overall territory is suitable for crop production. Consequently, the main speciality of the few state farms and over eighteen collective farms in existence has remained dairy farming.[1]

In the 1970s, the population of the entire Chernobyl raion, an area of 2,000 square kilometres, was only about 47,000. The capital, Chernobyl, had a population of just over 10,000, while sixty-nine tiny villages were scattered throughout the raion, dominated by intricate river systems, which feed the man-made Kiev Reservoir in the east. At this time, the large majority of the population was made up of Ukrainians, although there were evidently sizeable numbers of Russians, Poles and Belorussians. It is likely that the proportion of Ukrainians has declined since the construction of the power station, for which labour reserves were brought in from several areas of the USSR.

About twenty-eight kilometres to the north of Chernobyl lies the Belorussian border and Gomel oblast. Zhytomyr oblast of the Ukrainian SSR is located about fifty kilometres to the west, and Chernihiv oblast (also of Ukraine) is situated across the wide expanse of the Kiev Reservoir. From Chernihiv city, the Desna River bends its way southward to merge with the Dnieper just north of Kiev. The Dnieper connects with the giant reservoir as do several other smaller rivers that permeate these and other oblasts of Ukraine.

In 1970, the Soviet authorities decided that this remote region of rural Ukraine, which was not part of the traditional "breadbasket," would be suitable for the location of the republic's first nuclear power plant. Its distance from the major cities and the natural river systems rendered Chernobyl a suitable location. From the outset, a huge facility was anticipated. The new city to be constructed, Prypiat, was envisaged to grow to 150,000—200,000 residents at its peak—although this was an ultimate rather than an immediate goal.[2] The site of the new atomic town was close to the village of Kopach, some twenty kilometres north of Chernobyl and hence less than ten kilometres from the Belorussian border.

The authorities were optimistic about the construction from the outset. In 1971, before the building work had begun, the chief building engineer, M.I. Krasnikov, said that the speed of the construction would be unique, "shock work methods" would be applied, and that the workers perceived themselves as perpetrators of an experiment in technique.[3] The statement should not, however, be taken too literally. Such euphoria has been traditional at the start of construction of many Soviet nuclear power plants, but this has not always meant undue haste or, for the most part, adhering to the most moderate of schedules. And indeed, there were delays in building Chernobyl from the first. In March 1970, the Minister of Energy came to the site and made the ceremonial laying of the foundation stone. The second did not follow until about eighteen months later,[4] and it is not clear why the start of work was so protracted.

By October 1977, the first power block came on stream, a graphite-moderated 1000 megawatt reactor, similar to that developed earlier at the Leningrad nuclear power plant. Other units followed in 1978, 1981 and 1983. On average, each unit of twin reactors had taken seven or eight years to construct. The fifth reactor was scheduled to come on-stream in 1986 and a sixth in 1988. By 1986, Chernobyl accounted for 10 per cent of the USSR's total electricity-generating capacity, and was, along with Leningrad, the Soviet Union's largest nuclear power plant at 4000 megawatts.

Why was the plant constructed and what was its purpose? A former Soviet official was quoted in an American newspaper as saying that there was acute opposition to the building of Chernobyl station within the Ukrainian hierarchy. He commented that demonstrations occurred and that those in opposition could see no reason why an energy-rich republic like Ukraine needed to venture into the domain of nuclear power.[5] Some Western specialists have speculated on one possible reason for such development, namely the production of plutonium for the Soviet nuclear weapons programme at Chernobyl. The fact that it used a graphite moderator, unique in Ukraine since all the other plants under construction use water-pressurized reactors, seemed to support this theory, whose adherents have included nuclear physicist Jens Scheer of the University of Bremen and the *Financial Times*.[6]

It is possible that the weapons programme played a role in the early years of the station (certainly there is not enough evidence to refute such a suggestion). But the third and fourth reactors were clearly earmarked for electricity production. By the late 1970s the need for energy was more acute than at the beginning of the decade. In February 1986, *Radio Kiev* announced that the Chernobyl plant was part of the MIR system, i.e., the integrated grid system of the CMEA countries.[7] A 750-kilovolt transmission line runs from Chernobyl into Vinnytsia oblast, where it links up with another (older) line that runs westward from the industrial heartland of the Donbass to Vinnytsia and Albertirsha, near Hungary. The latter country imports about 25 per cent of its electricity and it is likely that the bulk of the supply now comes from the Chernobyl station.

Another factor that militates against the designation of Chernobyl as a weapons-manufacturing station is the invitation to officials of the International Atomic Energy Agency (IAEA) to inspect the plant in 1985.[8] In the event, the IAEA selected Novovoronezh, which used VVER reactors, but the invitation itself suggests that the Soviets had nothing to hide. Chernobyl is, and most probably always was, a plant intended solely for the production of electricity.

In February 1986, the magazine *Soviet Life* featured the Chernobyl plant, in an article by Maxim Rylsky entitled "A Town Born of the Atom." The intention of the article was to assure the reader of the utmost safety and viability of the Chernobyl station and the general satisfaction of the workforce and families with the situation in Prypiat. In one exchange, chief engineer Nikolai Fomin (who has since been relieved of his duties) declared that the plant was absolutely safe and that the plant's cooling pond, an area of twenty square kilometres, was being used for fish breeding. The reactor was housed "in a concrete silo" and possessed "environmental protection systems." Even if the incredible should take place, he added, automatic control systems would close down the reactor within seconds. The plant was said to be equipped with emergency core cooling systems and many other technological safety designs.

Boris Chernov, a 29-year-old steam turbine operator, maintained that fears of nuclear plants were unjustified:

I wasn't afraid to take a job at a nuclear power plant. There is more emotion in fear of nuclear power plants than real danger. I work in white overalls. The air is clean and fresh. It's filtered most carefully. My workplace is checked by the radiation control service. If there is even the slightest deviation from the norm, the sensors will set off an alarm on the central radiation control panel.

If the plant was safe, Prypiat was depicted as an idyllic city, home of over thirty different Soviet nationality groups who could see the outlines of the plant from their apartment windows. A young population, average age twenty-six, was facing a bright future, and only minor irritants were cited, such as a shortage of daycare facilities and nursery schools, employment for women, and shortage of garages and parking lots for cars.[9] Retrospectively, such comments appear somewhat inane, but even at the time, they ran counter to the very different picture of Chernobyl that emerged from Soviet accounts. *Soviet Life* is oriented primarily toward an American audience. Soviet citizens, on the other hand, may have been aware that some very serious problems had emerged at the nuclear power plant over the years that appear to have been even more serious than those at the other Ukrainian nuclear facilities.

Even in the 1970s, when it was one of only two Ukrainian plants under construction (the other was Rovno), it proved difficult to obtain a large enough reserve of qualified workers. V.T. Kizima, the head of the building department at the station, said in the summer of 1974, that workers had been given the task of bringing the first reactor on-stream by December 1975, but that the schedule was unrealistic because there was now a shortage of labour. Work had begun to outgrow the labour force, and the latter was predominantly unskilled. A need had arisen, he continued, to shift the emphasis from quantity to quality, so that the standard of work did not lag behind the speed of the construction. At the present time, Kizima stressed, "There is a definite gap between the levels of construction and the quality of that construction, and that imbalance must be redressed."[10]

One of Kizima's dilemmas — it was presented in the Soviet press as an asset — was that there was no middle management at the plant. Although the budget of forty million rubles per annum would have allowed for the assembly of a substantial team: four or five building management squads; four chief engineers and at least four deputies; Kizima was given direct and total control over the entire operation. Because the schedule for work had been stepped up in 1974, Kizima was obliged to seek help from subsidiary building organizations in other districts. A rapid pace was said to be necessary because of the past failure to adhere to the plan target, and all ancillary projects such as roads, communications (perhaps even the construction of apartments in Prypiat) were put aside so that the building of the plant could take precedence.[11]

Measuring work in terms of the amount of money expended, Kizima noted that whereas in April 1974, total outlay had been 2.7 million rubles, the May figure was 3.57 and for June, over 4 million rubles would be spent. At the same time, there was a concomitant expansion of the workforce, from an initial 2,200 to over 5,000, and a projected 7,000 by the end of the year. Most of these workers had no previous experience of work at a nuclear power plant. Kizima himself had worked previously at a "traditional power plant," which was "of no value" in the building of an atomic plant. Workers were said to find the surroundings congenial, yet many had no permanent accommodation, which can hardly have been satisfactory to young families relocated from other regions.[12]

By 1976, there were clear indications of a dissatisfied workforce. Chernobyl workers were losing ground in their "socialist competition" with the Kursk nuclear power plant (over the border in the Russian SFSR), largely because of the losses in working time incurred. During the winter months, Chernobyl nuclear power plant lost 12 per cent of every working hour. Within two years, the workforce had doubled from the 5,000 noted above to 10,000. Over half of the workers were "young people" and the problem, according to a Soviet source, was how these youngsters could occupy their leisure hours.

They could, it noted, spend their time usefully, go to the library, the theatre, the sports hall or the school, but some of them were falling into bad company, "and beginning to drink." Many arrived late for work, and took excessive breaks during the day. Overall, a lack of labour discipline and a high turnover of personnel were adversely affecting work at the plant. The local authorities were paying more attention to better social facilities, something "they should have done from the beginning." Yet there was a lack of subtlety about the nature of the recreation. After a long day at work, the average worker had little inclination for "intellectual films and books," and his tastes ran rather to lighter affairs.[13] There is a contrast here between the affirmed intention: completion of work on the first reactor of a major nuclear plant requiring dedicated and skilled work; and a young, frivolous labour force, frustrated by the shortage of housing and lack of amenities, bored into alcohol consumption in remotest northern Ukraine.

Subsequently, *Radianska Ukraina* observed that the authorities' main concerns about Chernobyl were twofold: the qualifications of the people involved in the building of the plant; and their state of mind. Because work had fallen so far behind schedule, shock workers were dispatched to the site from other areas, people with experience of completing major projects in a short space of time. Some of the processes were transformed to "factory conditions" to raise efficiency. The first reactor itself required about 520,000 cubic metres of monolithic concrete, 273,600 cubic metres of reinforced concrete, 38,000 tons of metal construction and 100,400 cubic metres of bricks.[14]

The workforce approached its enormous task in the late 1970s with a "trial-and-error" attitude. If problems arose, they had to be resolved without "ready answers." Thus workers had to learn how to make high alloy steel, and Kizima described Chernobyl as "the first university of atomic construction at which [the workers] themselves had to discover the solutions to problems."[15] To some extent, this should not be taken as a major drawback. In a new industry, the workforce has to be built from scratch, and Chernobyl was one of the earliest Soviet commercial nuclear power plants. Yet the labour force appears to have been unstable, workers left and others, with less experience, replaced them. If it was a university, it was hardly one that turned out a skilled product after a certain tenure—the students were leaving the campus at too frequent a rate. However, upon the completion of the first reactor at Chernobyl, as we have seen above, the more experienced Chernobyl workers were transferred to other nuclear plants at which the building work was just beginning. The "teething problems" therefore reoccurred with every new task, every new generating unit.

A Soviet emigre now living in Israel, Boris Tokarasky, was involved in the building of the Chernobyl plant before leaving the Soviet Union in 1978. He maintains that the reactor design and the management at the plant were "dangerously deficient in technical standards," that Soviet turbines and piping at the nuclear power stations are identical to those at the coal-fired power stations, and lack the sophistication required at a nuclear power plant.[16] While the statement seems plausible, it is counteracted to some extent by Kizima's comment above that his own experience at a traditional power plant counted for nothing at a nuclear station, i.e., Soviet officialdom does seem to have been at least aware that one could not simply transfer the experience and the system used at one facility directly to the other. This is not to say that they did not do so, however.

In the 1980s, the difficulties with the labour force have remained unresolved. In fact, they appear to have worsened throughout the decade. In July 1985, the First Party Secretary of the Prypiat urban party committee said that serious attention had been paid to the cadre problems at the Chernobyl nuclear plant, particularly among the construction workers:

The insufficient regard of the leaders for the task in hand, their low level of professionalism, low level of labour and production discipline led to frequent shortages and consequent non-fulfillment of plans.[17]

Later in this same year, the chaotic situation in supply procurement was noted by *Radianska Ukraina*, and in the article a party secretary pointed out that the document for supplies had been returned up to seven times for corrections to be made.[18] Thus labour and supply problems seem to have gone hand-in-hand.

In March 1986, the durable Kizima was interviewed in *Vitchyzna*, and the journal duly noted that during the course of his work on the fourth reactor and the building of residential and recreational facilities at Prypiat—for evidently he supervised both aspects—he had been awarded the Gold Star of the Hero of Socialist Labour and other awards. The article made clear how much responsibility had fallen directly on Kizima during the work, which corroborates Kizima's earlier statement that he had had to carry out his labours without a middle management. He faced some major tasks. For example, the designated number of workers at Chernobyl was based on figures for a hydroelectric station and proved to be a huge underestimate. Kizima had been obliged to take his problem to the Ministry of Power and Electrification in 1975, but the necessary funds for the expansion of the labour force had not been made available until 1980.

The builders, on the other hand, were said to receive their plans piecemeal, and worked on one section at a time without having any clear conception of the overall plan or the future structure. Consequently, during the construction of the fourth generating unit, a dead zone developed in an area for which no plan had been made available. Only "at the last moment" was a plan on hand relating to building in an adjoining unit.

Nevertheless, Kizima decided that, labour shortage, partial plans or not, progress could be accelerated, and early in 1984, he went to Moscow to "persuade" the Minister of Power and Electrification of the USSR (Neporozhny) to cut down the schedule for the completion of the fifth reactor at Chernobyl from three years to two (1987 to 1986). The argument for this revision of plan ran as follows: Kursk nuclear power plant had begun construction a year ahead of Chernobyl and had three reactors; Smolensk began a year later and possessed only one. Yet Chernobyl had four reactors, and clearly was a more model facility. If a fifth could be completed a year ahead of schedule, about 20,000 tonnes of coal could be saved, which clearly was a major factor given the stagnation of output at Ukraine's Donetsk coalfield. Therefore, said V.T. Hora, the chief engineer, "Kizima cut the red tape in building the fifth reactor by pushing the schedule forward by one year, thus compelling the bureaucrats in Moscow to complete the necessary documents faster than usual.''[19] The full implications of this move soon became evident in the Soviet press.

On 27 March 1986, an article by Liubov Kovalevska in *Literaturna Ukraina* gave a detailed account both of the equipment/supply difficulties and the very low working morale at the Chernobyl station. Kovalevska's position was not revealed, but some Western writers have speculated that she may have been an official at the plant. What is clear is that she had a detailed knowledge of the situation and few qualms about airing her complaints in public. The result was a graphic description of an ailing construction, a building, moreover, that would have been described as "lagging" in Soviet terms in any industry.

According to Kovalevska, the nuclear power industry was developing particularly rapidly in Ukraine. But work on the fifth reactor at Chernobyl was being plagued by shoddy workmanship:

> The building site should be an uninterrupted production line of work on the basis of the strictest adherence to correct building techniques. This is precisely what is lacking. The problems of the first energy block were passed on to the second, from the second to the third, and so on. But together with this they expanded, "became overblown" and there were a huge number of unsolved problems. At first, these problems were discussed with interest, with firm self-confidence, then they aroused indignation and later, desperation: "How long," they asked, "are we to continue talking about the same thing, and what is the use of all this talking?"[20]

The reaction of the authorities to this dilemma was, as we have seen already, to cut down the time for the completion of work on the fifth reactor from three years to two. And yet the project co-ordinators, the suppliers and even the construction workers were said to be quite unprepared for this change of schedule. The Zuk Hydroproject Institute did not provide the financial-budgetary documentation in time to order the necessary reinforced concrete and building metal. Consequently, the assembly organizations did not receive their construction orders until the latter part of the year, and assembly work was said to be disrupted. The chaos that resulted reportedly had a profoundly adverse effect on the individual worker:

> The disorganization of production weakened not only discipline, but also each individual's sense of overall responsibility. The inability and even unwillingness of engineering-technical staff to organize the work of the brigades resulted in a slackening of standards. One began to notice

"fatigue," the wearing out of equipment, machines and mechanisms, shortages of instruments, power tools, etc. In a word, all the defects of the construction process—which unfortunately are typical—also made themselves evident in extreme forms.

Again, the authorities' response to this situation, the seriousness of which can hardly be overestimated, was to put forward a highly unrealistic target expenditure of 120 million rubles on construction work for 1986, which, the author noted, was significantly higher than the previous maximum outlay of 70 million rubles. Kovalevska pointed out, however, that not all the fault lay with the planners and workers: the suppliers were falling short of required standards too. For example, in 1985, of the ordered 45,500 cubic metres of prefabricated concrete, 3,200 were missing, and from the amount that did arrive, 6,000 metres were found to be defective. As a result, work stopped. Later in the article, the author became sarcastic, but even more forthright:

Equally helpful to the builders of Chornobyl atomic energy station last year were the suppliers of metal structures, who undersupplied by 2,359 tonnes, and what was delivered was largely faulty. This included 326 tonnes of fissure sealant for the nuclear fuel waste depository, which arrived in a defective state from the Volzhskii metalworks. The same plant was partially responsible for defects in the manufacture of girders for the machine hall. The Kashira metalworks sent nearly 220 tonnes of faulty columns for its assembly....Among the shoddy producers one frequently encounters the Prydniprovsky works of the Union Atomic Energy Construction Industry association which is the main supplier of farraginous concrete for the Chornobyl atomic energy station.

Kovalevska's article was perhaps the most serious criticism of the Chernobyl plant to appear in the fifteen years during which construction work had taken place. It is a matter for conjecture whether a crisis had finally come to a head, or whether the article represents merely the most recent of a long string of attacks on the nuclear station. In retrospect, it appears to have been prophetic, but it should perhaps be read differently, as an indication that Soviet plans were unrealistic, and moreover, were alienating an already unhappy workforce.

The human factor cannot be dismissed in the light of what followed, and given the fact that similar situations existed at other Soviet nuclear power plants. To a Western observer, who visited a Soviet nuclear power plant in May 1986, the RBMK plant looked shoddy, a "tin shed," poorly constructed, and with what appeared to be an inadequate concrete shield over the top of the reactor, through which steam was being emitted.[21] The observation rings true. Chernobyl was a badly built edifice, with a demoralized workforce.

The Accident

At the time of writing, the full explanation of what occurred at Chernobyl in the early hours of Saturday, 26 April 1986, had not been provided by the USSR.[22] Initially, Western writers condemned the Soviet authorities outright for what they felt were inadequate safety standards. The London *Daily Telegraph*, for example, called Chernobyl a "textbook example" of how the environment and safety of Soviet citizens is sacrificed to the needs of the Soviet economy, and said that in the cutting of corners, safety precautions are thrown "out of the window."[23] Similarly the Economist Intelligence Unit's *Quarterly Energy Review* said that the accident was probably a result of skimping on routine maintenance work and the overworking of plants as a result of delays in bringing new nuclear power plants on-stream.[24]

Gradually it became clear from comments by Soviet officials that while the above two comments may not have been totally off the mark, something more intricate, if not more sinister, had occurred. As early as 7 May, Borys Shcherbyna, the Donetsk native who was the first head of the Government Commission investigating the disaster, said that the cause might be a combination of totally implausible events.[25] Three days later, the Hungarian MIT correspondent, Laszlo Fazekas, who had been touring the Kiev area, remarked that the fourth reactor block had been shut down for maintenance work and was operating at an output of only 200 megawatts (or 20 per cent of its total generating capacity) at the time the accident occurred.[26]

On 12 May, one of the designers of the Chernobyl plant, the prominent nuclear physicist Ivan Emelianov, hinted at a failure of the plant's emergency cooling system:

> At the critical moment, the automatic protection system quickly brought the reactor to a sub-critical state and the chain reaction of fission was stopped. But before the reactor stopped, the chain reaction with uranium-235 had generated a considerable number of radioactive fragments. If, following an accident, *the system intended to consume the concomitant heat is idle*, the reactor gets heated and this may lead to a conflagration of a quantity of the graphite inside it.[27]

Emelianov also referred to "an almost improbable coincidence of chances." At this stage, the world became aware that the plant was working at low capacity, that some odd combination of events had occurred, which led to a surge of power from the low capacity to 50 per cent in a matter of seconds. Emelianov had noted that the power increase was well above the maximum allowed for in the plant's design, which was a 16 per cent increase over a ten-second period. Finally, it appeared that there may have been a problem with the plant's cooling system.

In mid-May, the Deputy Chief of the USSR Committee to supervise safety in the nuclear power industry, V. Sydorenko, who had been in the Chernobyl region within a matter of hours of the accident occurring and was thus presumably in a good position to know something about the event, stated that experiments were being carried out on the fourth reactor when the accident happened. The explosion had taken place when the heat output rose from 6 [not 20] per cent to 50 per cent in 10 seconds. "The accident took place at the stage of experimental research work."[28]

A fuller explanation was given by Boris Semenov at a press conference held in Vienna on 22 May. Questioned about Sidorenko's statement, Semenov denied that any experiments were being carried out on the reactor, but postulated that there could have been experiments on the power-generating turbine, after which the power surge overtaxed the reactor's cooling system. The water in the reactor turned to steam, which combined with the zirconium alloy protecting the fuel rods to form hydrogen. The latter exploded, releasing approximately 10 per cent of the fuel in the reactor core, or about 18 tonnes of radioactive matter. He reiterated that the chain reaction had stopped automatically, but said that the top of the reactor had been damaged and the reactor building blown open.[29]

With the dismissal of several officials at the station on 20 July 1986, it became apparent that in contrast to the indications of Sidorenko and Semenov's remarks, the experiments taking place at the time of the accident had not been authorized. *TASS* revealed on this date that that the accident had been caused by "a series of gross breaches of the reactor operational regulations" by workers at the station. The experiments had taken place at a time when the reactor had been sidelined for planned repairs at night. The managers and specialists at Chernobyl, continued *TASS*, had not prepared for such an experiment, nor had they agreed to its taking place with the proper authorities, "although it had been their duty to do so." The experiments, moreover, were conducted without either the proper supervision or the necessary safety measures. A picture thus emerges of irresponsible officials, of human error once again. It is surprising, however, that Sidorenko, at least, was unaware that such experiments were unauthorized, several weeks after the event.

Yet the experiments themselves were not called into question. There was no indication from the CC CPSU Politburo report that Soviet leaders were concerned about the *fact* of an experiment. Evidently it would have taken place at some point in the near future. So perhaps a more fundamental question is: why were experiments to be carried out on a commercial reactor? A representative of Atomic Energy of Canada Limited (AECL) stated that the idea of performing experiments on a reactor in public service would not be entertained in Canada. Yet not only was this the case at Chernobyl, but also the experiments were unauthorized and carried out at night on a weekend, when many officials would surely have left the premises. This raises serious doubts about the level of work discipline at the station that had already been thrown into question by articles in the press.

Given the circumstances, and the Soviets' stake in the future of the plant, in addition to the industry as a whole, it was not surprising that Soviet officials began to refer to ''human error'' as the likely cause of the accident. Emelianov had said as much, and even Western journalists joined in the chorus, postulating among other theories that an operator had made a basic error in positioning the fuel rods that control the reactor.[30] Gradually, however, Soviet officials admitted that the causes were more complex. Evgenii Velikhov, the vice-president of the Academy of Sciences of the USSR and the chief scientist involved in investigating the disaster was interviewed on CBC's *The Journal* in late June 1986 by Mike Duffy, and in somewhat strained English explained the nature of the problem:

DUFFY: Have you been able to determine what actually caused this accident?

VELIKHOV: It is complicated, the official report is not yet finished. Our goal is to finish in August.

DUFFY: Do you believe it was human error?

VELIKHOV: Of course. I think it is [sic] some combination of human problem and technological problem...if such an accident is possible, it is a problem of technology.

Velikhov implied that any plant should have had sufficient fallback safety systems to prevent a loss of cooling in the event of a human error — since humans are always fallible, technology should be infallible. A more pessimistic view might be that technology likewise is never foolproof.

The view of a Canadian expert is that the psychology of the Soviet authorities pertaining to nuclear energy may be quite different from that in Canada. The basis of Canadian thinking is that although severe accidents may have a low probability of occurrence, it is necessary to design reactors to mitigate the consequences of these accidents. Accidents are not impossible, and it is necessary to continue to investigate severe accidents to ensure that safety systems are adequate. In contrast, the Soviet authorities—and this has also been demonstrated earlier—believed that a major accident was impossible. It was unimaginable and therefore should not be imagined. Only with this in mind can one conceive of the unprecedented build-up of nuclear power in the USSR. It also explains, not the accident causes themselves, but why the Chernobyl incident proved to be such a major disaster, why a combination of minor factors led to the inferno that ensued after the reactor overheated.

Whether or not the emergency cooling system worked at Chernobyl, it was not adequate to contain an accident from above. Emelianov pointed out in an interview with the Italian Communist Party newspaper *Unita* that the ''Western method'' of placing a protective cap over the reactor was substituted in the Soviet Union by installing a water basin underneath it to gather and condense the vapours in case they were expelled.[31] This emergency cooling system may or may not have failed after the reactor became overheated, but clearly there was no sprinkler system from above—thus, a ''dry accident'' occurred. Nor was there any sort of device for controlling the hydrogen that was produced, causing the explosion.

The production of the hydrogen from the zirconium alloy and steam, while never a probability under normal circumstances, might have been seen as a perennial possibility, given the alloy's properties. Zirconium, which is used in both the USSR and the West to protect the uranium fuel contained in the reactor core, has a very low neutron capture rate, a high melting point and is fairly corrosion resistant. Found mainly on the beaches of the east and west coasts of Australia, zirconium has a great affinity for oxygen, with which it reacts at very high temperatures to form zirconium oxide and hydrogen (it does not react with graphite, however). An AECL official has estimated that a fire might break out at a temperature of about 1,000 degrees celsius, although the chances for such a fire would increase greatly at 1,200–1,400 degrees. The normal operating temperature of the reactor, according to the manual cited above, is 280 degrees.[32] Chernobyl's emergency cooling system was geared for some lesser event, such as a break in a 90 millimetre pipe. According to one specialist, ''it had safety features, but not adequate standards.''[33]

Some features of the accident can be pieced together. The turbine generator was kept running after the reactor had been shut down.[34] There was overpressure in the tubes inside the steel vessel, followed by a hydrogen explosion, after which a crane fell onto the core of the reactor. The pressure sucked water out of the core leading to the dramatic rise in power from 6 or 7 to 50 per cent of capacity. There may have been local hotspots well above the 50 per cent, however, which contain the full explanation for what occurred. The pressure tubes led through a steel liner into a room with no containment. There was thus a direct path to the environment for the thermal plume. At the same time, the graphite may have sucked in oxygen from below leading to a further reaction, and other gaseous products may have been emitted with the destruction of the spent fuel storage base.[35]

Three features rendered this sort of accident unique to the Soviet RBMK:

1. The use of graphite, as a moderator (which caught fire).
2. The absence of water to contain radioactivity. Soviet provisions were for problems to be resolved from below the reactor, but the weakest point may have been at the top.
3. The lack of an adequate containment structure. The Soviet claims that the building was strong enough to withstand an airplane crashing on the roof are not corroborated by those Western observers who have toured an RBMK facility (in most cases that at Leningrad, which is the official RBMK showpiece for foreigners).

In addition to the above, there were evidently some problems involved with controlling the stability of the reactor. The amount of enriched uranium used for this purpose had been increased from 1.8 to 2 per cent, but control rods (some of which are for safety purposes and some to control the reactor) were changed frequently to keep the reactor stable.

Following the explosion, the major task was to quench the fire that engulfed the fourth generating unit, which soon had reached the roof of the adjoining third reactor. The nuclear plant's own fire brigade unit arrived within two minutes of the accident, according to the record of G.V. Berdov, a Deputy Minister of the Ukrainian Ministry of Internal Affairs. He also wrote that "over fifty" fire engines arrived from Kiev and the surrounding areas.[36] The weekly *Ekonomicheskaia gazeta* confirmed that "there were soon fifty fire-fighting teams on the scene."[37] On the other hand, while it is likely that most of the fire engines came from Kiev itself, it is implausible that they were at the site in time to play a major role in the fire quenching. As with the arrival of the evacuation buses on 27 April, the road from Kiev to Chernobyl nuclear plant is a lengthy one, and about two hours would have been necessary for the journey. According to *Pravda Ukrainy*, however, the fire was "practically out" (i.e., the flames had been extinguished) by 5 am.[38] For the most part, therefore, from approximately 1.30 to 3.30 am, the local firefighters bore the brunt of the task—a fact that is also corroborated by the high casualty rate from radiation sickness among local firefighters. Further, a recent Soviet publication in the West stated that the fire was fought *by twenty-eight men*,[39] i.e., the "outside" teams did not arrive until the worst danger was over.

Another Soviet account states that Chernobyl fire crew consisted of only fifteen men. Station fire-chief Major L.Teliatnikov, interviewed from Moscow's Hospital No. 6, said that he realized at once that "a crew of fifteen men could not cope with the fire."[40] The first two teams at the scene were those of Lieutenants V.P. Pravyk and V.N. Kibenok, and reportedly it was upon getting into his fire truck that Pravyk radioed "Call Number Three," in response to which every fire truck in Kiev oblast was obligated to rush to the scene.[41]

The small squad therefore faced an immense task, and appear to have carried out their duties with selfless abandon. A Soviet television broadcast of 6 June said that the firemen did have geiger counters to measure radiation levels, but that these simply "went off the scale." In short, the firemen were aware of the dangers they faced in the early hours of Saturday, 26 April.

The fire raged in at least five different areas. The main tasks were to prevent the flames from engulfing the third generating unit and spreading into the cable canal networks that extended throughout the entire power station. Teliatnikov sent one division to protect the machine room, while two others, "at the cost of incredible efforts," protected the third reactor block. Here the threat emanated from the blaze on the roof of the machine room. Firemen A. Petrovsky and I. Shavrei were ordered to climb the turntable ladder to put out the fire there. They remained for 15–20 minutes—"it was impossible to stay there any longer"—put out the fire and were then "picked up" by first-aid units.[42] In Shavrei's words, they were, by this time "in bad shape."[43]

Teliatnikov and others were fighting the blaze at a height of 71 feet, where the instrument department and the main brunt of the fire was located. Part of the roof over the reactor had already collapsed, load-bearing structures had warped and a "scorching hot torrent" of burning bitumen was surrounding the firemen on all sides. Poisonous and dense fumes made breathing difficult and reduced visibility, and there was the constant threat of sudden jets of flame accosting the firemen.[44] After the roof fire had been extinguished, the main fire began to threaten the engine room, which contained tanks of oil, and the cable shaft that linked all the units of the nuclear plant.

The firemen soon began to suffer the costs of their immense task. Kibenok witnessed his colleague Volodymyr Tishchura "writhing and squatting," then "Mykolai Vashchuk swayed and fell flat on his back." Vasylii Ihnatenko was close by and was to be another casualty.[45] Teliatnikov himself and sixteen others were badly injured during the course of events, but Teliatnikov's team, which included Kibenok and Pravyk remained at the scene until 5 am, by which time the main fire was said to have been extinguished.[46] Again, the main task had been borne by a few men. Soviet sources have revealed that the following played major roles (and many of the names were later to adorn gravestones in a Moscow cemetery): L. Teliatnikov, V. Pravyk, V. Kibenok, N. Vashchuk, N. Titenok, V. Tishchura, I. Shavrei, A. Petrovsky, S. Legun, M. Nychyporenko, V. Ihnatenko and V. Pryshchepa.

Radiation

The effects of high-level radiation on the human system are well known. Since the discovery of radiation by Roentgen in 1895, several data bases have enabled scientists to deduce maximum permissible levels: survivors of Hiroshima and Nagasaki; early uranium miners who did not have adequate protection; and medically irradiated groups. Experiments on mice have added to the data available and statements have been drawn up (inter alia) by the International Commission for Radiological Protection (ICRP) and U.S. National Academy of Sciences. The effects of low-level radiation, however, are more open to conjecture. Some scientists assume that the danger remains proportional to the dose received, but it is fair to say that this conclusion has not been accepted universally.

The effects of radiation can be divided into two groups: non-stochastic and stochastic. The former group comprises such categories as: sickness and death; cataracts; birth defects; loss of fertility; loss of fitness; and accelerated ageing. All the above increase in severity according to the dose of radiation received. The second group is made up of genetic disease and cancer, for which the severity is not necessarily proportional to the dose. A very small amount, well below the permissible levels may be a cause of cancer. For the stochastic group, it might be assumed that the chances of contracting cancer would increase with higher doses, but the cause and effect syndrome is not clear.

Deliberations of scientists have brought forward the conclusion that if 10,000 people were to receive 1000 millirem of radiation, then during their lifetime, one of those 10,000 would get a cancer—and this cancer may arise between 4 and 40 years after the dose of radiation is received. Of 10,000 Canadians, approximately 2,300 might be expected to die of cancer in the 1980s under normal circumstances. Exposure to 1000 millirems of radiation would increase cancer expectancy from 0.2300 to 0.2301, i.e., by a negligible and virtually immeasurable figure.[47]

The average exposure from nuclear power plants in Canada during the course of a year is 5 millirems, as compared to the natural background level of 80 millirems a year, and even this figure assumes a hypothetical individual living at the boundary of a nuclear power plant.[48] On average, an individual is exposed to between about 120 and 300 millirems per year, and the amount may vary according to such diverse factors as the number of medical X-rays one receives during a year, or the number of long-distance air flights, or living in an area where granite predominates, such as France's Massif Centrale, where levels are well above average.

The ICRP "safety factor" permits the workers a ten-fold leeway by setting the maximum permissible norm at 5000 millirem, that is to say acute non-stochastic effects of radiation are considered to be in the area of 50,000 millirem (50 rem) per year. For the public, a further factor of ten renders the maximum permissible dose 500 millirem annually. In short, the most that the public should in theory have been exposed to would be about 100 times less than a dosage that would have an acute effect. This factor is of significance in examining the effects of Chernobyl and explains to some extent why Western analysts were maintaining that levels were well above maximum permissible norms at the same time that the Soviet and Polish authorities were emphasizing that the levels in areas away from the site were "absolutely harmless" (which is not to justify the banality of the latter remark).

The dose received according to the above categories assumes that the entire body is irradiated. Thus if only half the body is irradiated with the hypothetical 1000 millirems noted above, then 2000 millirem would have to be accumulated to have the same effect. To put matters into perspective, one can cite the scale of radiation levels listed by AECL in 1983:[49]

5 millirem: the maximum radiation received at the boundary of a nuclear power plant;
100 millirem: the normal background level from natural sources of radiation at sea level;
500 millirem: maximum level for the member of the public in Canada;
10,000 millirem: no observable effect if given instantaneously;
1 million millirem (1000 rem): if received instantaneously by the entire body, would cause illness and resulting death within a few weeks.

One could add an intermediate category to the above, namely about 325,000 millirem would kill 50 per cent of those exposed, according to a Western source.[50]

One might conclude from these figures that in the course of daily life, nuclear power plants are relatively harmless. After Chernobyl, anti-nuclear activists have claimed otherwise. The main question, however, is how much radiation the people at the plant and in the vicinity were exposed to.

Concerning harmful effects, three main areas are affected by high-level radiation, namely gastrointestinal parts, the blood cells and the brain. An intense dose of radiation above 1000 millirem to the entire body in a brief period would destroy more blood cells than the body could replace, resulting in death within a brief time.[51] The so-called "Derived Release Limits" for iodine-131, however, have been been set at 500 millirem per year, and it is clear that iodine was the most prevalent of the isotopes released from the thermal plume at Chernobyl.

In the case of the accident, even with the wind that carried the radioactive cloud northwestward into Scandinavia, an umbrella pattern of dispersal emerged that limited the fallout of the nuclides. As the particles returned to the earth's surface, those that are gaseous could have been inhaled, while those attached to the soil could be inhaled and the nuclide released into one's system. Iodine-131 is a mobile and soluble element that emits gamma rays that can penetrate the human organism to the bone. It attacks the thyroid, which uses iodine in the natural course of life to produce thyroxine. If the iodine is in a gaseous form, it can be taken up by plants, which are subsequently consumed by animals. It can pass through the plant back into the soil and be taken up, in turn, by other plants, creating a vicious cycle. However, the half-life of iodine-131 is only 8.04 days. Consequently, in the Chernobyl region, it would not have affected the grain harvest, and the Soviet authorities soon began to conclude that leafy vegetables and milk could be stored until safe for human consumption. In the interim, iodine could be administered to the population in order to saturate the thyroid and prevent the ingestion of the radioactive iodine in the atmosphere.

Among the other significant elements of the hundreds released into the atmosphere by the Chernobyl disaster were caesium-134, caesium-137, strontium-89 and strontium-90. Strontium is known as a "bone-seeker," and decays to form another unstable element yttirem. Penetrating the human body, it travels to the bone, and then the blood stream, and tries to replace calcium. As with iodine, a principal area for concern is milk ingestion, and inhalation is another principal means of penetration. Strontium-89 has a half-life of 50.52 days, while strontium-90 has a half-life of 29 years. Caesium-137, which emits gamma rays and is most dangerous at the ground level, has a half-life of 30.17 years. Aside from the iodine therefore, other long-lasting elements, potentially harmful, were released into the atmosphere. A feature of the aftermath of Chernobyl has been the release—from the Soviet and Polish governments in particular—of information about the impact of iodine-131, such as the amount permissible in milk, and the total ignoring of measurements of those isotopes that were released in smaller quantities like caesium, strontium, tellurium and others that are still potentially dangerous.

On 10 May, *Radio Kiev* stated that a medical examination of the workers "directly involved" in the accident (but presumably not the firemen) did not reveal a high level of radioactivity. The main part of the radioactive flow, it continued, was contained in short-lived radionuclides, half of which were made up of the isotope iodine-131. "Soon, a lowering of radiation occurred in the 30-kilometre zone." "At the moment of the accident," it declared, "the highest level of radiation in the 30 kilometre zone was 15–20 millirems per hour." By 5 May, this figure had declined to 2–3 millirems and on 8 May, the reading was 0.15 millirems. The normal hourly background in the area was said to be from 0.005 to 0.0025 millirems per hour. Thus the figure of 15–20, represented at most an increase of 4000 times the normal rate. If one applies the ICRP standards to this rate, one can conclude that a dangerous level had not been reached, according to the maximum figure given by Kiev radio. At the same time, however, the maximum permissible hourly dose of radiation for a member of the public, on the basis of ICRP standards, had been exceeded by over thirty-five times. Only by 8 May did radiation fall to an acceptable level. (The radiation standards in the USSR are derived from ICRP recommendations).[52]

To take the Soviet figure of 15–20 millirems per hour at face value, however, would be erroneous. It has been shown above that the radiation level to which the firemen were exposed was too high to be measured on the geiger counter. U.S. doctor Robert Gale also informed that 35 of the severely injured patients had been subjected to doses of over 800 millirems.[53] Moreover, *Radio Kiev* referred to the 30-kilometre zone as a whole. It is clear therefore that the 15–20 millirems per hour referred to the average level, and that the dosage in the immediate vicinity of the reactor must have been many times higher than this.

In reporting the affair, however, some Soviet authorities constantly played down the dangers, creating the illusion that away from the damaged reactor, radiation posed few if any dangers to the populace. For example, Iu.A. Izrael, the Chairman of the USSR State Committee for Hydrometeorology and State Control, was reported as stating in mid-May that "an increase in the level of radiation in any area is now ruled out," that the doses accumulated in the 30-kilometre zone were within the norms acceptable to the IAEA, but that the decision to evacuate the area had been taken for the people's safety and health. A "slight increase in the level of background radiation" had been observed in a number of cities in Ukraine and Belorussia. In Kiev, however, the level of 0.3–0.4 millirems per hour "posed no threat to health." A "slight increase" in the background radioactivity in Poland, Romania and Scandinavia "also posed no danger."[54]

A colleague of Izrael, Nikolai Kozlov, the Deputy Chief of the State Committee, went a step further than Izrael and informed foreign journalists on 16 May that iodine and "other elements released" had half-lives

133

of 3–14 days, which at best was only a partial truth and a misleading statement because most would have assumed that he was speaking of all the elements emitted.[55] Similarly, L.P. Feoktistov, the Deputy Director of the Kurchatov Atomic Energy Institute of the USSR Academy of Sciences, noted on 17 May that the level of radiation in the city of Kiev was less than 5 rems per year, and could be compared to the effects of a dental X-ray. Clearly, he pointed out, this "presented no danger to man."[56]

Feoktistov presented his report on Soviet television. Thus he was unlikely to say anything that might cause consternation among the viewers. Whether the level at Kiev was 0.3 millirems per hour more than three weeks after the accident may not have been of much relevance. Gradually, a more worrying picture emerged from Soviet and East European sources. On 17 May, the day before Feoktistov dismissed the dangers of radiation before Soviet citizens, *Izvestiia* contained the following warning:

> The distance from the headquarters of the clean-up campaign to Chernobyl atomic energy station takes only about 10 minutes. But this distance is fraught with unseen danger. The point is that even here, right next to the station's damaged power unit, radiation is not evenly spread across the entire area. It looks as if it has "pock-marked" the soil. Nothing in some places, quite dense in others.

Implied in this statement is that in other areas too, the radiation was not distributed evenly. A reading taken in one location might not apply a few metres further on. And it would not have been possible for the Soviets— or anyone else in the same predicament—to monitor levels at intervals of only a few metres. An IAEA statement issued after representatives of the Vienna agency had visited the accident scene revealed that the Soviet authorities had distributed potassium iodine tablets widely both inside and outside the 30-kilometre zone,[57] which suggests that in reality the Soviets were not entirely convinced of the wisdom of their asserted beliefs about the radiation levels outside the declared "danger zone."

At the end of June, the Polish Ukrainian-language newspaper *Nashe slovo* revealed that the Polish authorities, at least, were not certain how much radioactivity had been released by the accident:

> The radiation was dispersed by the fire and the smoke and was of a very high level until the crater was plugged. Aerosols, small particles...were carried some distance by the wind. The radioactive smoke contained graphite, uranium and particles of metal. The smallest aerosols created a cloud which moved toward Belorussia, Poland and Scandinavia. *But this*

cloud contained only a small part of the radioactivity that escaped from the plant. The total amount is unknown since which particular part of the radioactivity was trapped in the core, and what percentage escaped has not been determined.[57]

Nashe slovo stated frankly that the people who were working at the Chernobyl plant at the time of the accident (i.e., not the firemen), were exposed to a level of radiation of "several hundred rems," adding that a dose of 400–450 rems is fatal in 50 per cent of cases. Taken at face value, this could be taken to mean that about half the people working at the station at the time of the accident, in addition to those who subsequently fought the blaze, could be expected to succumb to the effects of radiation. The newspaper also stated that "some of the radioactive aerosols must have fallen" into the Kiev reservoir.

In late May and June, it was evident that while radiation levels in the Chernobyl region were abating, they were still a cause for concern around the station. On 1 June 1986, Moscow television stated that the levels around the Chernobyl station were still high enough to necessitate strict controls in the immediate area and that workers' shifts were sometimes limited to only a few minutes, depending on the amount of radiation in the work area. At the same time, the radiation level was said to be falling constantly, at a daily rate of about 5 per cent.

Early in June, several "dirty spots" of high radiation were found outside the 30-kilometre zone in Belorussia, forcing the evacuation of over 60,000 children from the southern regions of Gomel Oblast.[58] Almost simultaneously, the optimistic Iu.A. Izrael was claiming that the radiation situation had now stabilized as a result of the physical decay of the elements and extensive decontamination work. Izrael said that people were already returning to certain areas of the 30-kilometre zone—"the first evacuees will be returning literally today or tomorrow."[59] A curious situation thus emerges, with areas being evacuated and repopulated in the same zone at the same time. The main point to be gauged from these contradictory remarks is that it was not possible to estimate accurately the amount of radiation in a given area. One wonders also about the logic of returning evacuees to a decontaminated zone given that factors such as a change in wind direction or heavy rainfall could still change circumstances so quickly.

A reading of Soviet accounts leads to the conclusion that radiation fallout was most severe in the immediate area of the damaged reactor. Consequently, one would have expected the level of radiation to fall correspondingly once one crossed the Soviet border into Poland. In northeastern Poland, however, concentrations of iodine-131 in milk reached a maximum of 1,720 becquerels per litre in the period 28 April to 2 May.

This was some 72 per cent higher than the maximum level permitted for children in Poland. Atmospheric readings, however, were well below permitted norms.[60] Nevertheless, one would have expected radioactivity in Chernobyl raion to have been higher than this level.

Similarly, with the longer lasting caesium isotope, high levels were found in food samples from Bavaria and East Germany at the end of June. In Neuruppin, north of Berlin, the caesium level had reportedly risen by 200–500 per cent since the disaster, while soil samples taken in Bavaria revealed levels of up to 40,000 becquerels per kilogramme. Fish caught in Sweden also contained about ten times more than the allowed level of radiation.[61] Another source has confirmed that there was "an unexpectedly high ratio of caesium in the fallout from the Chernobyl reactor" and postulates that this could have an adverse effect upon Ukrainian agriculture.[62] There was little indication from the Soviet side that these sort of levels abounded.

Boris Semenov, the Deputy Chairman of the USSR State Committee for the Utilization of Atomic Energy, thus noted in June that the situation was "normalizing." According to readings from the Ostior station, he added, the radiation level sixty kilometres from the station was 0.14 millirems per hour on 2 June. In the northwest of the USSR, the dosage rate was said to be the same as the natural background.[63] In fact, the Soviet authorities have thus far given no information about the highest levels of radiation, the amount of contamination of food, or whether there was ever any danger posed to the citizens of Kiev by the disaster.

Casualties

No country would have been adequately prepared for a nuclear disaster of the magnitude of Chernobyl. The fact that the accident occurred at night and on a weekend made the Soviets' task even more difficult. In the Soviet case, there was a marked shortage of experts in medical radiology at the local hospitals, which was, according to Oleg Shchepin, the First Deputy Health Minister, "a serious gap in the training of people in our health system."[64] As a result, the most seriously injured persons at Chernobyl were sent to Moscow's Hospital No. 6, a speciality hospital geared to treat victims of a nuclear accident.

The head of that hospital, A. Guskova, a graduate of Sverdlovsk Medical Institute, was reportedly "on the hotline" to local medical establishments by 1630 hours on 26 April, i.e., some fifteen hours after the accident occurred.[65] It may therefore have taken some time before either the seriousness of the injuries was ascertained, or before it was realized just how unprepared the hospitals of northern Ukraine were for an event of such magnitude. By 1800 hours, two of the staff from No. 6 Hospital, T.T. Toporkova and G.D. Seledovkin, were on a flight to Chernobyl to assess the situation and to decide which of the victims could be dealt with locally, which should be sent to hospitals in Kiev (the intermediate cases) and which to Moscow (the worst cases). Within a twenty-four-hour period, the two doctors and others had evidently examined 1000 persons.[66]

On 27 April, the 206 seriously injured victims were flown to Moscow. According to the IAEA report, which followed some two weeks later, they consisted predominantly of nuclear plant personnel and firefighters, all of whom were affected by radiation from the first to the fourth degree. Two persons, as already noted, were said to have been killed outright, Volodymyr Shashenok and Valerii Khodemchuk. Western writers and politicians were alternately sceptical and incredulous concerning the official total of two dead. Matters were not helped by the erroneous UPI report of 2,000 victims of Chernobyl at the outset, which confused many Western readers, but was commendably ignored by the Associated Press. The photographs of the damaged reactor, the size of the plant and of Prypiat (well over 25,000 residents) and the reports from U.S. satellite sources of a possible "second meltdown" (*Time* magazine entitled its issue on Chernobyl "Meltdown"), suggested a much higher casualty figure. U.S. arms negotiator Kenneth Adelmann and Secretary of State George Schultz were among those who refused to take the figure of two dead seriously.

With regard to this figure—and the Soviet propagandists have not been slow to respond to the mistaken impressions that were given in the West—it should perhaps be said that it was at best an interim total. There could be little justification for Western speculations, but there was also little point in reiterating the figure ad nauseam in the Soviet news agencies, since there was never a possibility that two dead could represent an ultimate or even immediate casualty figure. *Glasnost* (openness) from the east would in turn have led to accuracy in the West. Instead, a human disaster became grossly politicized from the beginning.

In the first days after the accident, about 100,000 people were examined for radiation sickness, and 230 medical teams, made up of 1,300 doctors, nurses and dosimetrists carried out the inpsections. Rest facilities were provided at the Lesnaia Polnaia clinic-sanatorium, near Chernobyl, where less seriously injured workers were taken for check-ups and to recover for periods of up to two weeks.[67] About half the patients hospitalized in the first days after the accident were reportedly discharged by 12 May.[68] Two days later, Mikhail Gorbachev announced on Soviet television that nine people had now died and 299 were still in hospital.

The disaster took on an international aspect early in May, when U.S. bone marrow specialist Robert P. Gale took a team of specialists to Moscow to assist Soviet specialists with bone marrow transplants for the most seriously affected victims. There is some conjecture concerning how Gale, a graduate of the New York State University in Buffalo, found himself in Moscow. Writing in the *Los Angeles Times*, on 29 May 1986, Gale stated that "On May 1, the government of the Soviet Union *asked me* to come to Moscow to aid Soviet physicians and scientists." Gale wrote that he responded "immediately," and with the aid of Occidental Petroleum chairman, Armand Hammer, assembled a team that included his colleagues at the UCLA Medical Center, Paul Terasaki and Richard Champlin, and an Israeli doctor, Yair Reisner.

Writing in the same edition of the *Los Angeles Times*, Armand Hammer stated that as soon as the Chernobyl disaster became known in the West, Gale "offered his services" to the Soviet government to provide bone-marrow transplants: "He knew that he must ask the Soviets directly, for this was a private offer and Moscow had already rejected U.S. government offers of assistance." Gale reportedly called Hammer, who dispatched a telegram to Gorbachev (Hammer has been on good terms with every Soviet leader, from Lenin onward). Hammer also used his personal funds to purchase $600,000 worth of medical supplies, even though the Soviets had offered to pay for these supplies.

Within a forty-eight hour period, Gale and his team were in Moscow. Whichever account is accurate, the West was now provided with a voice in Moscow, although Gale was careful never to say anything that might offend the Soviet authorities or to compromise his work in Moscow.

Gale and his team participated in the main medical work carried out after the disaster, the monitoring and care of the most critically injured patients. Guskova has emphasized that Gale's team supplemented an already proficient team assembled at the hospital rather than provided expertise lacking in the Soviet team. She pointed out that her colleague, A.Ia. Baranov, had already performed six bone-marrow transplants by the time Gale arrived. They were assisted

> by the Cardiology Institute, with its highly developed biochemistry, the Haematology Institute, with its blood service, and the Epidemiology and Microbiology Institute, which produces special diagnostic preparations for us and assesses the concentration of anti-infection drugs.[69]

Each patient was provided with a 24-hour individual doctor and nurse. Ten junior physicians were duly promoted and became heads of sections overnight, while the doctors "had to acquire new skills." Disputes occurred with the American doctors, but according to Guskova's account, Gale and his colleagues usuallly acceded to the Soviets' demands: "I am sure that we would behave similarly in his clinic."[70] According to Champlin, the American team worked 12–15 hours daily, but they and their Soviet colleague were at first limited by a "critical shortage of technology" and with equipment that was over twenty years old. Most of the equipment used was of Western origin, he noted, and were not manufactured in the USSR. Also, the Soviets simply were not equipped to handle a large number of casualties. Blood cells were still counted under a microscope as opposed to an automatic counter, which took about 30 seconds while the long hours at the hospital were necessitated by the frequent breakdowns of machinery, and in order not to delay crucial transplants.[71]

According to Robert Gale, attention was focused initially on the 35 patients who had suffered the most substantial doses of radiation. Altogether, he noted on 6 June, 80 people had suffered serious radiation exposure in the accident, and 400–500 were admitted to hospitals. Thirteen patients had received bone marrow transplants by this date, of whom eight had died, while six were given injections of fetal liver cells.[72] Another Western doctor at the hospital, Michael McCally, a professor of clinical medicine at the University of Chicago, said that bone marrow transplants had been carried out on those who had received over 500 rems, "the lethal dose." He also stated that the death toll at the end of June stood at 26, but added that 200 workers, doctors and firemen remained "critically ill."[73]

Champlin provided a harrowing portrayal of the agonies experienced by one victim, of blisters on the face and mouth, ulcers across the body, red "weeping" skin burns. The victim's membranes that lined his intestines "had eroded" and the patient suffered from "severe, bloody diarrhoea."[74] Nevertheless, the Soviet *Vremia* television programme, which was to receive heavy criticism from *Pravda* for the way it covered the disaster, saw fit to interview Chernobyl firemen from the hospital in early June. Teliatnikov, Petrovsky and others were shown, clad in pyjamas, gaunt and losing hair, all declaring in their turn that they felt "fine" and were anxious to return to their homes. Guskova also stated optimistically that "It seems to us that we have the wherewithal to enable them to return to life and work."[75]

Gale himself was interviewed by CBC Television (Toronto) in late June, and described the major damage to the bone marrow caused by the massive doses of radiation, and to the skin from the graphite fire which had been burning at a temperature of up to 5000 degrees celsius. The burns from radiation, he explained, could have resulted from touching radioactive particles or from external radiation. As the blood count had begun to fall, the patients became susceptible to bleeding complications and infections. He felt, however, that the victims were aware of what had happened to them and that they had knowingly entered a zone of high radiation: "They were firefighters or physicians trained to deal with a nuclear reactor accident."[76]

Teliatnikov and Petrovsky evidently recovered enough to visit the graves of several of their colleagues at the Mitinskoie cemetery on the outskirts of Moscow later the same month. Having been discharged from Hospital No. 6, they were to spend "several months" at the Luniovo sanatorium near Moscow. Altogether, eight of the firemen were discharged.[77] As for those buried in the Moscow cemetery, most were Ukrainian firemen from the Chernobyl plant, including Titenok, Ihnatenko, Kibenok, Pravyk, Vashchuk and Tishchura.[78] The death of Kibenok, one of the first two firefighters on the scene was the occasion for some eulogies in the Soviet press at both the national and (Ukrainian) republican level. A young man of twenty-three, he left a pregnant wife, and died fifteen days after receiving a "powerful dose" of radiation during the fight against the fire.[79] Other "known" victims include a plant worker, Lelechenko, shift leader at the fourth generating unit, A. Akimov, and an operator, A. Kurguze. In late May, Professor A. Pozmogov from a Kiev clinic informed Western correspondents in Zurich that ten of the patients there "were in a worrying condition.[80] It is by no means clear therefore that all the victims died in Moscow.

In addition to the firemen, the patients in Moscow included two first-aid workers and other plant personnel. No helicopter pilots had suffered as a result of their missions, and only two residents of Prypiat had suffered from radiation sickness, both of whom had wandered onto the plant site. By 2 August, the total number of victims had risen to thirty, a total that, in Gale's view, would not get much higher.[81]

To estimate the final death count from Chernobyl would be to indulge in speculation. Instead, one should merely keep in mind the postulates cited above, namely that the number of victims rises in proportion to the dose of radiation received. The number of deaths due to non-stochastic categories is not likely to exceed one hundred. The grey area remains the stochastic category, future cancers and genetic diseases. Gale and his colleagues decided to monitor the progress of about 100,000 people from the general area around Chernobyl over the next decades. Only over a period of about forty years therefore can an accurate estimate of the number of Chernobyl victims be offered.

As for the Soviet medical crews, they appear to have performed as best they could in an unprecedented situation. While Soviet hospitals are clearly less well equipped than those in the West, there is little reason to doubt the ability and competence of Soviet doctors such as Guskova and Baranov. Guskova's grimaces and sharp temper provided a more realistic depiction of what was happening at Hospital No. 6 than *Vremia*'s somewhat macabre interview with firemen suffering physically from radiation sickness, prematurely aged and balding.

The Evacuation

Soviet reports about Chernobyl abound with descriptions of heroism, selfless behaviour and arduous work. A less than glorious chapter in the history of the event was the evacuation—or rather evacuations, since there were several—of citizens from contaminated areas. The political aspect of the process will be discussed in Chapter Seven. Suffice it to say here that regional and Moscow officials did not agree about the danger of radiation contamination of local residents. Indeed, the local officials appear to have had little conception of the danger even after the dramatic events of the early hours of 26 April. Only a few hours after the main blaze was extinguished at the nuclear plant, and firemen had suffered enormous doses of radiation of between 400 and 800 rems, Prypiat children were making their way to school. They recalled trucks hosing down the streets with water, and were given a warning from their teachers to stay indoors when they returned home, to change their clothes and warn their parents.[82]

An even more astonishing example emanates from Belorussia. Emphasizing the "normality" of day-to-day existence, an article in *Pravda* cited a soccer game being played between two raion capital towns, Khoiniki and Bragin. The two towns represented several villages that had been evacuated and could be said to be dangerously close to the evacuation zone. A first game was played on 13 May, with a return match on 18 May.[83] On 7 June, *Izvestiia* warned residents of Kiev not to play soccer on the beaches because the sand being kicked up might be contaminated. Subsequently, the city of Gomel, about 150 kilometres to the north, i.e., much further away from the danger zone of 30 kilometres around the Chernobyl plant, was evacuated. Why were the soccer games permitted? The most charitable explanation is that the local authorities were ignorant of the dangers, just as the Prypiat urban party committee allowed schoolchildren to attend school on the same day that the disaster occurred.

Reports about the entire evacuation process are riddled with contradictions and inaccuracies. At best, it can be said that it proceeded with "organized chaos." Neither the Soviet authorities nor "welcomed" outsiders could obtain a clear picture. For example, following the visit of Hans Blix and other IAEA officials to the area, the IAEA issued a statement that up to 48,000 people had been evacuated from Chernobyl and other locations within a 30-kilometre radius.[84] On the following day, a correspondent of Budapest radio stated that he had been touring the Kiev area and had learned that just over 90,000 had been evacuated from the Chernobyl region.[85] In both cases, the spokespersons derived their information from Soviet officials. Since the Budapest correspondent's estimate turned out to be closer to the truth, one must question why the IAEA was given inaccurate information.

A similarly distorted picture emerges over the first hours after the accident. Once the Chernobyl accident had become the centre of a polemical conflict—predominantly between the USSR and the U.S. media—the Soviets began to manipulate some of the information provided about the evacuation. Clearly one of the weakest "chinks in the Soviet armour" was that the population of Prypiat and those farms in the vicinity of the nuclear plant had not been evacuated on the day of the accident. Consequently, the *date* of the initial evacuation began to be omitted from Soviet accounts. A June edition of the weekly *Ekonomicheskaia gazeta*, for example, noted simply that following the accident "The inhabitants of Prypiat were evacuated in 1100 buses in less than three hours."[86] There was no indication in the article that this signified three hours *on the following day*.

Concerning the dates of the evacuation, it was by no means clear initially which areas had been evacuated. As a Western analyst points out, the First Deputy Minister of Foreign Affairs of the USSR, Anatolii Kovalev, stated at a Moscow press conference to which foreign journalists were invited, that a danger zone of 30 kilometres had been demarcated and that the Chernobyl region had been evacuated within thirty-six hours of the accident. The logical deduction was that all those within the danger zone had been evacuated within that time period. Yet two days later, the Ukrainian premier, O. Liashko, stated that only people within a 10-kilometre zone around the damaged reactor had been moved, and the evacuation of the larger zone, which included the town of Chernobyl, did not take place until a week after the accident.[87]

In fact, there does not seem to have been any strict adherence to the 30-kilometre limit, which prominent officials such as Velikhov never took very seriously.[88] Villages such as Dytiatky (25 kilometres from Chernobyl), Strakholissia (30 kilometres from Chernobyl) and Hornostaipil (25 kilometres) were never evacuated, reportedly because radiation levels in these villages were normal.[89] The same report indicated that there were a number of other villages also exempted from the general exodus. On the other hand, once out of the danger zone, farms were not always totally abandoned. Members of the Prypiat state farm, which is almost adjacent to the atomic power plant, were evacuated on 27 April, but subsequently returned to the farm to pick up their equipment "which was first decontaminated."[90] The danger zone was evidently not considered so dangerous by some of the residents, particularly in cases when they were being asked to help with farming work in their new location and lacked the tools to do any work.

Although some villages were never evacuated, the current official figure of 92,000 evacuees still seems too low when all the information is taken into account. Evidence suggests that the number of evacuees, migrants and those who simply fled the scene, when added to the thousands of schoolchildren sent to summer camps (there were reportedly 60,000 removed from Gomel oblast in June alone), and the large numbers of citizens who were leaving the city of Kiev, it is plain that the minimum overall figure is well over half a million, and that the 92,000 refers to the Ukrainian total alone rather than the 30-kilometre zone. We are informed from Soviet sources, for example, of the following figures:

a) Ukraine [92,000 total]:
Prypiat workers relocated to other nuclear plants and various industries—20,000;
Evacuation to the town of Poliske (population 40,000)—23,000;
Evacuation to Bragin raion, Belorussia—7,000;
Evacuation to the town of Makariv—6,000;
Evacuation to Ivankiv raion—approximately 7,000;
Evacuation to the village of Zahaltsi (Borodiansky raion)—2,200;
b) Belorussia:
Gomel raion residents evacuated from the southern part of the raion—26,000;
Evacuation of Belorussian children—60,000;
Overall total: 151,200

To the above must be added the removal of over 250,000 schoolchildren, pregnant mothers and mothers with children under seven years of age from Kiev and other cities,[91] and the evacuation of many citizens from the city of Gomel, concerning which very little information has been made available, but which appears to have been substantial. One report, for example, revealed that hundreds of evacuees from Gomel were living in Tallinn, Estonia in July.[92] Not all the moving occurred at the same time, and not all those in transit have been recorded in official accounts, but there is verifiable evidence in most cases that the migration occurred.

From official sources, one can deduce the following. On 27 April, in the early part of the afternoon, residents of Prypiat (population 25,000-40,000) and farms located close to the reactor, such as the Prypiat state farm mentioned above, were evacuated. Initially, the destination was Poliske raion. Over 23,000 evacuees and the Prypiat urban party organization were relocated in the town of Poliske itself,[93] which is located about 50 kilometres west of Chernobyl. The total would appear to account for the majority of the population of Prypiat.

Following the visit of Ryzhkov and Ligachev to the site of the accident, the town of Chernobyl itself and other villages within the 30-kilometre zone were also evacuated between 3 and 6 May. A week later, the school term was brought to a close, and children from Kiev and other areas were sent to summer camps on the Black Sea and other locations. Further north, on the Belorussian side of the border, an evacuation of 26,000 residents of the southern part of Gomel oblast took place at approximately the same time. Initially, it appears that the Belorussian police forces had arrived in Khoiniki, and established their provisional headquarters in the town. This was considered expedient because several villages in the raion were located about 5–7 kilometres "as the crow flies" from the damaged nuclear reactor. It was "soon obvious" that evacuation was essential as radiation levels had risen sharply. Consequently the police supervised the evacuation of three "densely populated raions" of Gomel oblast, Belorussian SSR, namely Narovlansky, Khoiniki and Braginsky. Pregnant women and mothers with children under seven were evacuated first, and then villages on the Ukrainian border, such as Radan Massany, Chamkov and Ulasy were cleared.[94] Almost half the Gomel evacuees were schoolchildren, mothers with small children and pregnant women.[95]

Once evacuated, the Soviet authorities tried to accommodate people in environments similar to their own: state farmers were relocated on other state farms, collective farmers on collective farms. Yet the process was not always well organized, there was stubborn resistance from certain quarters, families were separated, even party officials had difficulty locating each other, and while it is reported widely that the evacuees were welcomed everywhere with great warmth and compassion—and there is no reason to doubt the veracity of these statements—there was often simply no room to put the new arrivals.

On 9 May 1986, *Sovetskaia Belorussiia* commented with regard to the evacuation that "the suffering is being shared by all." In some cases, it was quite plain that the residents had no wish to leave. According to the Second Secretary of the Chernobyl Raion Party Committee, V.M. Mernenko, the district centre—Chernobyl—was evacuated after the town of Prypiat in order to give the party committee "time to prepare the population psychologically to leave."[96] This might well be construed as yet another explanation for the delay in moving people out of the danger zone, but this is not to deny that residents were unhappy with the idea of moving.

The First Party Secretary of Khoiniki Raion Committee in Belorussia, D.M. Demichev, also said that "it was very difficult for peasants to leave familiar places...there were tears. Some old women hid in cellars...."[97] Some Prypiat residents simply refused to move. Two elderly women, A.S. Semeniaka and M.I. Karpenok (aged 85 and 74 respectively), hid during the evacuation and were only discovered more than a month after the accident. Some old men found their way back to their apartments—presumably from their new locations—"nobody knows how," and were discovered by a patrol. Upon apprehension, they reportedly protested, "We will not leave—who will feed the geese and chickens?"[98]

In the town of Bragin, the residents were not moved, even though the radiation level apparently warranted an evacuation. According to *Izvestiia*, it was not easy to accommodate 7,000 people in districts that had already received over 30,000 people from the evacuation zone. Accordingly, a hazardous operation ensued, in which the town was decontaminated without moving the residents.[99]

Despite the delay in commencing the evacuation, people were hustled onto buses so quickly that they left valuables behind in their apartments. In some cases, residents may have been led into believing that their sojourn away from home would be short. According to police colonel A. Vasiliev, patrols in Chernobyl were maintaining round-the-clock protection of peoples' property. Valuables, money and documents had been left behind, electric lights and fridges left on. The patrols had begun operating on 1 May, at first on foot, and then later in armoured cars.[100] The fact that police were asked to fulfill such a duty reveals that the local authorities, like some of the residents, did not anticipate that Prypiat was to become a "ghost town."

In mid-May, *Pravda* bitterly attacked the way in which the evacuation had been carried out. Some party and government leaders, it noted, were not meeting their responsibilities in caring for evacuees. Some evacuees could not find "their loved ones." Mothers with small children should by now have been at summer camps and rest homes in the Black Sea area, but there were still "hundreds of mothers with small children" in Poliske raion, which was evidence of the "bureaucratic indifference" of some Kiev workers.[101] In June, *Pravda* reported again that "some local authorities were not doing what they should to help evacuees." Letters sent to the newspaper revealed that one evacuee was still searching for relatives, while another, a woman, who had been relocated at Brest, located on the border with Poland, could neither obtain a job nor manage to send her children to a summer camp. *Pravda* this time laid the blame on local press, radio and television, which should have been keeping the public better informed.[102]

When the evacuees arrived at their new destinations, conditions were difficult. The farms in the south of Bragin raion in Belorussia were said to be prosperous, and the residents there were used to having amenities such as clubs and canteens. Those in the north had no such comforts. Moreover, the new accommodation was lacking in showers and baths, rooms were hot and dusty. Camp-beds were "brought in by the hundreds," but sometimes could not be squeezed into the small houses. Evacuees slept on the floor, sometimes two to a single mattress, eight or nine people crammed into one house.[103] Under such circumstances, the evacuation could not be a smooth or easy process. It constituted a major upheaval in peoples' lives, and perhaps could not have been otherwise even with the best of organization.

What is not clear from most Soviet reports is how widely the evacuees were dispersed. Having removed most of the small children and mothers to resorts, the remaining citizens who could not be accommodated on the already overcrowded farms in the southern and western parts of Kiev oblast, were scattered across the Soviet Union. An elderly woman wrote to *Pravda* complaining that she had been unable to obtain a pension in Saratov, near the Kazakhstan border, because she had left behind her pension book. Another letter came from Baku, Azerbaidzhan.[104] Some Chernobyl residents were found employment in Lithuania, where they were provided with apartments for which local residents may have been waiting from five to ten years.[105] Others were sent to Tula (200 kilometres south of Moscow) and Tallinn.[106]

It remains unclear how many of the evacuees will return to their homes, or when the area will be deemed safe for children to return to the vicinity. Soviet accounts have spoken of the eventual return and "normalization" of the contaminated zone, but it was stated in *Pravda* that the authorities had no intention "of gambling with people's health." *Pravda* also revealed that preparations were in hand to build barns to accommodate thousands of cattle moved from the 30-kilometre zone, and to construct 10,000 winter homes for evacuees,[107] which suggests that the stay in safer regions was to be prolonged. Changes in the levels of radiation, which allegedly led to the July evacuation from the city of Gomel noted in *Trud*, make it difficult to declare any zone "safe" with absolute certainty.

Not since the Second World War has the Soviet population been forced to evacuate a region. At that time, the enemy was visible and his intentions soon clear. After the Chernobyl disaster, regional officials seem to have underestimated the enemy, and then, once the decision to evacuate had been made, little care was taken to keep families together, and many evacuees were unaware of the length of time that would be spent away from home.

The Public

As was seen in Chapter One, the city of Kiev did not suffer its peak levels of radiation until about nine days after the accident because the wind was blowing the radiation cloud northwestward rather than to the south. The May-Day parade and an international bicycle race took place on schedule, in the case of the former, the day before representatives of the CC CPSU Politburo accompanied Ukrainian party leader, V.V. Shcherbytsky, and Kiev oblast First Party Secretary, V. Revenko, to the accident area. From the beginning concerning the city of Kiev, there was a contrast between some of the announcements on the local Kiev radio and statements in newspapers, Moscow radio and Soviet news agencies. Also, the rumour mill, which is always evident in a major city, was operating from the beginning (some of the stories that circulated in the days after the disaster are outlined below).

On 9 May, Revenko was quoted in *Pravda* as stating that the "radiation situation in Kiev presents no grounds for fear." The cities, streets and plazas were, as always, swarming with people, while factories, offices and shops were in full service. Milk and vegetables, he stated, were being checked carefully, first at their point of origin and again at the Kiev entry point. Children and breast-feeding mothers had been sent to summer camps purely as a precautionary measure. Revenko's assurances were supported by *Radio Moscow*, which also made the bland statement that "life is normal in Kiev" and interviewed a number of foreign students, all of whom testified that there was nothing untoward happening on the geiger counters they had with them.[108]

On 8 May, the day before the radio's broadcast, however, its Kiev counterpart had repeatedly issued a warning from the Ukrainian Ministry of Health. A change in the wind direction and in its strength had "increased the level of radioactive contamination in the city." The level was not believed to be dangerous for health (the fact that no figures were provided suggests that it may have been above recommended norms, however), but some prophylactic measures were advised. These included the obligatory closing of all windows to prevent the spreading of radioactive dust and the contamination of food; shaking dust off clothing after being in the street, wiping floors, carpets and furniture with damp rags; daily showers and washing hair with soap or shampoo.[109]

The ministry also advised that residents should not venture into the countryside, beaches, gardens or allotments unless absolutely necessary. People should refrain from staying in open spaces. They should consume vitamin-rich food, especially Vitamin B, drink more fluids and stop eating spinach, sorrel and salad. Food should only be bought from stores at which radiation levels were inspected regularly. All wells used for drinking and household purposes were to be covered. A virtual crisis situation evidently was imposed for a few days, although the precise times and duration of the emergency are difficult to gauge.

Thus on 9 May, Ukrainian Minister of Health, A. Romanenko, addressed Kiev citizens on *Radio Kiev* and informed them that the radiation level had gradually lowered "and is now within the norms recommended by international and national organizations," i.e., it had been above those norms previously. At the same time, he said that dust was now the main enemy and that alertness and surveillance were essential in the days ahead.[110] A Ukrainian newspaper, however, noted two days later that a change in the wind direction had seen a rise in the level of radioactivity in Kiev and the surrounding areas. Residents were instructed to take precautionary measures, to wash themselves and their homes regularly, to stop smoking and drinking alcohol, not to eat greens, and to keep children from playing on the ground.[111]

By 15 May, at least 250,000 school-age children had been evacuated from the city and sent to summer camps. Many citizens evidently attempted to follow them, jamming the Kiev railway station. *Pravda* referred sardonically to "whirlpools of hysterical, selfish individuals" at Kiev railway stations,[112] while a Moscow-based newspaper noted that the demand for railway tickets in Kiev "just before the May-Day holiday" had been so great that officials had to open eight extra ticket counters at the station, cut staff lunch breaks and extend working hours.[113] The authorities had laid on extra trains for people who wished to leave the capital "for fear of radiation exposure."[114] The Reuter correspondent in the city observed that 200 people were waiting hopefully for airline tickets at the Aeroflot office, while residents were reporting problems in obtaining rail and air tickets out of the city.[115]

According to *Pravda*, Kiev residents had been "vulnerable to false rumours from the West" because they had not been given complete information about the accident initially.[116] A driver in the city told the London *Times* correspondent, Christopher Walker, that residents had been living "a normal life" for over a week after the accident until they were suddenly warned on 5 May that they should be taking numerous precautions, similar to those mentioned above.[117] The sudden increase in radiation levels seems to have alerted the authorities belatedly to the dangers. Whereas foreigners who had been removed from the city on the advice of their governments in the first days after the accident were assuring the Western media that reporting in the West had been grossly exaggerated,[118] the real danger to Kiev came later, approximately 9–10 days after the accident.

Consequently, many Western sources were prepared to accept at face value Soviet statements that no danger existed, from the logical deduction that radioactive decay would by this time have reduced the radioactivity from its original high level. But in fact, the reverse had occurred, as Soviet sources admitted in their warnings to the population. Robert Gale, who cannot in any respect be considered an anti-Soviet source, stated in early June, one month after the warnings had been issued, that the radiation levels in Kiev were still approximately 15–30 times above normal.[119]

There is firm evidence that some panic occurred in the city, Soviet reassurances notwithstanding. As early as 11 May, A.M. Kasianenko, the Deputy Minister of Health of the Ukrainian SSR stated that uncontrolled use of medicines (presumably iodine) could be dangerous, and that "There is no truth to the rumour that alcohol is useful against radiation."[120] Guskova herself was questioned by an *Izvestiia* interviewer about these allegations:

> *Interviewer*: We have been to Chernobyl. There is much talk that vodka helps in the event of irradiation.
> *Guskova*: No! I have been telephoned from Kiev. People were asking about red wine and vodka. Alcohol deceives and prevents a person from correctly understanding his own condition.[121]

The rumour-mill evidently did not abate and the question found its way into *Sovetskaia Rossiia* in June 1986. People were reportedly "holding serious" discussions about a round-the-clock trade in vodka that had begun in Kiev, how vodka prices had been reduced drastically, and how all the drivers working in Chernobyl were being given table wine because "as people are saying, wine and vodka work well against radiation."[122] Evidently the widespread belief was that as radioactive substances may have entered the body and were difficult to remove, various alcoholic combinations should be tried to remove the "poison" from the patient.

This and other related questions were examined in mid-July by the newspaper *Komsomolskaia pravda*. In an article entitled "Radiation: Myths and Reality," it refuted the "persistent public notion" that garlic or alcohol could be useful against contamination. "How can alcohol," it asked, "working chemically on the organs, influence radiation, acting physically on the level of cells?" It also emphasized that taking pharmacy iodine as an antidote, "as some people have been doing," was not only pointless, but also dangerous. Nor could people contaminate others, because they had absorbed far too little radiation for this," the newspaper pointed out.[123]

In the Chernobyl region (and also over the border in Poland),[124] several pregnant women decided to have abortions after the disaster, while others fled from the area before the authorities organized the evacuation. Many even left Ukraine altogether, considering it dangerous to be anywhere in the same area as the damaged reactor. *Pravda Ukrainy* condemned their actions, albeit in the mildest fashion, as being "without foundation."[125] Vladimir Sokolov, the Kiev correspondent of *Radio Moscow* expressed frustration with the "rumours and untruths" that were being aired about the station, about Chernobyl and about Kiev, which were pervading the Ukrainian capital: "What can one invent about Kiev, where normal life is still going on. . . . ?"[126]

Yet the rumours were not limited either to Kiev or Kiev oblast. Muscovites were also said to be frightened about radiation, to be asking for medical examinations. They were evidently afraid not only of contact with people who had come from Chernobyl, but even about touching their belongings or handling letters from the area.[127] On 2 May, *Pravda* revealed that it had received letters from readers opposing nuclear power. Valerii Legasov, a Deputy Director of the Kurchatov Atomic Energy Institute of the USSR Academy of Sciences, was told by the newspaper's interviewer: "I will not keep it a secret that our editorial mail includes letters which express a negative outlook toward atomic energy."

To the "official rumours" noted here can be added "unofficial rumours," which also may not always have been based on fact, but nonetheless circulated in the city of Kiev and only added to the consternation and fears of citizens.[128] There was said to be widespread discontent about certain features of the evacuation: pre-school children, for example, had not been evacuated to summer camps in mid-May. Further, while there were no restrictions placed on leaving the city, the authorities never on any occasion advised citizens that this would be a preferable course (this is clearly accurate). It was believed, on the other hand, that the Ukrainian party chief's grandson had been whisked out of the city on 27 April, implying that the danger was self-evident to the Ukrainian authorities. Further, all institutions of higher learning were permitted longer vacations so that academics could stay away from Kiev.

Posters reportedly were displayed at Kiev University warning people not to trust the statements of the authorities, and to get their children out of Kiev. These, it is said, were not removed for several days. About one million Kievans abandoned their jobs and left the city unofficially after the accident, but the authorities allowed people to return to their positions without reprisals, because officials were so sensitive about the Chernobyl affair. Although children had been sent to summer camps for 45 days, they would have been returned to their homes had not a prominent academician protested, securing a further 45-day extension. A letter from a Kiev resident to the West noted that the evacuated children would not be returned to the city until the leaves had fallen from the trees (i.e., the leaves contained radioactive particles).

As for illness, it is reported that virtually every sickness was being blamed on the Chernobyl disaster. Kiev relatives of Prypiat citizens were believed to have died after being contaminated by these same persons after the accident. A radiation specialist informed a Kiev resident that he had contracted a cataract of the eye that appeared to be a result of exposure to high levels of radiation. It was virtually impossible to obtain dosimetric apparatus to measure radiation levels. One citizen came to the West in search of a dosimetre because he had been unable to locate the machine in either Ukraine or Eastern Europe.

More verifiably, Kiev residents began listening to Western radio reports on a mass scale, even travelling as far as the Carpathian Mountains to circumvent the heavy Soviet jamming. Western broadcasts were taped and circulated. Even jammed broadcasts were taped in the hope that a technical means could be found to eliminate the jamming and still hear the original broadcast. At Kiev libraries, all literature on the U.S. Three Mile Island accident of 1979 was withdrawn from the stacks.

The general picture is of a dissatisfaction on a much wider scale than is apparent from Soviet reports. Clearly Kiev residents were very concerned about the accident, about their state of health, and about the confusing and often contradictory statements of the authorities about radiation levels, and what steps should be taken. Many citizens did not accept the Soviet statements that the level of radiation in Kiev had never been damaging to their health.

CHAPTER SEVEN

After Chernobyl

The Clean-Up Campaign

The profound psychological impact of the disaster upon the Soviet authorities manifested itself in various ways. Even while mobilizing every available force to assist in "eliminating the consequences of the accident," and while following the official line that Chernobyl had revealed a glimpse of what life might be like for the world after a nuclear conflagration, some Soviet writers also offered frank and profound thoughts on the disaster. They revealed that in some quarters, there was a feeling of helplessness in the presence of unknown forces unleashed by the Chernobyl disaster. Cited here is one example from a political writer of the Soviet government daily, *Izvestiia*:

In the past when fate took an unexpected turn, our forefathers used to say: everything is at the mercy of God. Without involving ourselves again in the argument about God, let us set out the evident truth of our time that everything is now at the mercy of the atom. And that applies to all mankind and all life on earth, whether the atom is used for military or peaceful purposes. We have been reminded about that from time to time by accidents in the United States, Britain and other countries. Another tragic reminder was provided by the accident at the Chernobyl atomic energy station, which suddenly entered our life—and not ours alone—and world politics.

For almost two weeks Chernobyl has been featured in newspapers and on the Vremia TV programme.[1] It occupies no less important a place in our thoughts and feelings.... Pictures taken from helicopters show the deserted settlement around the AES, the bright new apartment blocks, the straight and empty streets. Sensing the meaning of this picture as people of the nuclear age, we feel the unseen, silent and awesome presence of increased radioactivity. So this is what it is like, we thought, as we watched....

We are not the only ones thinking about Chernobyl. This word, until recently unknown to most people in our country, is now on the whole world's lips. On such occasions, you see once again how small the world is. In the modern poet's metaphor, "The world dangles in a string bag of latitudes and longitudes" containing almost five billion people gripped with fear at the threat of nuclear war to which has suddenly been added the danger of radiation from the Chernobyl atom....People are casting many wary glances at the sky and their surroundings—radioactivity needs no visa and has no respect for national frontiers.[2]

Amidst such reflections, the Ukrainian authorities began to organize personnel to deal with the enormous task of a burning graphite fire and a plant that was emitting a steady stream of radiation into the atmosphere through the yawning crater that was formerly the nuclear plant's roof. Reactors one, two and three required a skeleton staff to monitor the cooling down process after they had been switched off. A devastated area had to be decontaminated and immediate and urgent measures had to be taken to prevent the pollution by radiation of the Prypiat River that flows into the main water supply for the city of Kiev.

On 28 April, *TASS* announced the formation of the Government Commission to investigate the accident under the leadership of Borys Shcherbyna. The latter's role was never very clear. The *Novosti* report issued in the first week of May, but referring to events of 28 April, made it clear that at that time Shcherbyna was spending his hours between Prypiat and Chernobyl. Subsequently, however, he seems to have spent much of his time in Moscow, and left on-the-spot duties to subordinates, most notably to Ivan Silaiev and later (when Silaiev was replaced) to Lev Voronin. There is little doubt that Silaiev played a more active part in the clean-up work than did Shcherbyna, but since the latter was not seen in public after mid-May, he may have fallen ill, as was rumoured, and have been unable to play an active role.

When Shcherbytsky visited the accident area on 1 July, it was reported in the Ukrainian press (cited earlier) that he met with Vladimir Gusev, who by that time was head of the commission. Gusev's name had not been mentioned previously in the press, and rarely appeared afterward. The likelihood, then, is that the commission did have its base in Moscow and left the fieldwork to prominent officials who would then report to the chairman. Nevertheless, the chairman was clearly rotated. By the first week of August, Gusev had been replaced by Deputy Chairman of the USSR Council of Ministers, G.G. Vedernikov, who took part in a high-level meeting in Chernobyl on 9 August and was referred to as Chairman of the Government Commission.[3]

Much of the work, which in late-May involved over 1,000 personnel,[4] was left to the army and police forces. For example, the sealing off of the plant site, evacuation procedures and the mission to extinguish the burning graphite were placed under the command of the Ukrainian police, and the army. The decisions were made by the Prypiat and Chernobyl sections of the Ukrainian Ministry of Internal Affairs, which were temporarily amalgamated and renamed the Chernobyl-Prypiat city raion section.

As early as 26 April, it is reported that the Deputy Director of the Ukrainian MVD, G.V. Berdov, went to Prypiat and closed all the routes to the plant and to the town. Additional police squads arrived from Poliske, Ivankiv and Chernobyl to Berdov's headquarters in Prypiat. Shortly afterward, Ukraine's police chief, I.D. Hladush, arrived "and took over command." He worked out an evacuation plan and "persuaded the population that such a course was in their best interests."[5] To maintain a "forbidden zone," watchtowers were established, "similar to those along the border [of the USSR]." The perimeter encompassed by these watchtowers covered a distance of 107 kilometres. This artificial boundary was then patrolled constantly by the militia and subunits of the Ministry of Internal Affairs. The latter were put in charge of issuing passes to those who entered, and entry points were reduced to three throughout the entire zone.[6]

Because radiation levels in the zone bordered by the watchtowers were very high, the task force's headquarters was removed from the Prypiat building, and by 9 May, an underground bunker had been established in the Prypiat area as the new centre. Around this date, control of the zone, which had been entrusted to the police, was now given to the army, under Colonel U. Keleberda, who had the immense task of supervising and co-ordinating the work of the army, civil defence teams, and various government departments and ministries that were arriving at the accident site.[7] As for the deserted town of Prypiat, the job of protecting the rapidly vacated apartments, which contained in many cases both valuable and documents, was given to a subunit of the Kiev Security Guards department [*Vnevedomstvennaia*], under the command of Colonel D.D. Chaus. Guards were placed at the entrances to each apartment block on four-hour shifts.[8]

The headquarters of the Prypiat party organization was transferred first to the town of Chernobyl, and subsequently to the safer locale of Poliske. Oblast party leader Revenko made occasional visits to the accident site, but the Ukrainian Party leader Shcherbytsky played a more distant role. He made only three announced visits between 26 April and 1 July, and on the first occasion, he merely accompanied Ryzhkov and Ligachev to the accident site. By contrast, Shcherbytsky's Belorussian counterpart,

N.N. Sliunkov, reportedly flew to the contaminated zone in the southern part of Gomel oblast several times a week by helicopter.[9] There is nothing to suggest, however, that Shcherbytsky was taking a back seat or disinterested role in the clean-up. Arguably, since his role at the accident scene could have been little more than ceremonial, there was little reason for him to play a more active part. Sliunkov's involvement, however, was unusual, but the Belorussian situation was less clear-cut. Radiation levels there rose sharply shortly after the accident and the evacuation process was even more difficult than in Ukraine.

Over the next weeks, a Chernobyl task force was created. It was made up of military and police forces, firemen, physicians, coal-miners, metro-builders, transport workers, officials and workers of the Ministry of Power and Electrification, the Ukrainian State Motor Vehicle Inspectorate (GAI), officials of the Ministry of Transport Construction and numerous others. Fleets of trucks, buses and cranes were dispatched to Chernobyl nuclear plant, along with assorted equipment for decontamination, sealing and tunneling work. Images were evoked in the press of the "Great Patriotic War," and certainly the locale took on many of the aspects of a war zone.

As early as 26 April, the Soviet authorities evidently came to the conclusion that the graphite fire had to be extinguished from the air. The commander of the helicopter pilots that were to carry out this perilous mission was Air Force Major-General Nikolai Antoshchkin. According to the armed forces' newspaper, *Krasnaia zvezda*, he was ordered to "leave urgently for the town of Prypiat" on the evening on 26 April, but because of the necessary preparations that had to be made, he only arrived in Prypiat on the following evening. Assembling a crew that included several veterans from the Soviet war against Afghanistan, Antoshchkin began his mission on 28 April,[10] i.e., at least 48 hours after the accident had occurred.

The authorities had decided to shut off the flow of radioactivity from the air, thereby placing a "stopper" over the reactor. Large MI-8 helicopters took off from a meadow located about 14 kilometres from the damaged reactor, loaded with bags that contained dolomite powder, sand and lead pellets (the latter were to fill in any gaps that were left by the sand). Once at the reactor, they hovered in the air at a height of 270 metres, while their cargo was unloaded.[11]

Initially, there were two main concerns. First, total accuracy was essential as workers were still manning the neighbouring reactor units. To ensure that the pilots hit their target, a man stood on the roof "close to the damaged reactor"—obviously at great peril to his life, since at this stage there was no protection whatsoever from the radiation—and directed the pilots.[12] Ninety-three drops were made on the first day, followed by 186 on the second, "all with 100-per-cent accuracy" according to Soviet accounts.[13] About ten tonnes of sand were dropped between 28 and 30 April, but in Shcherbyna's view, it was like trying to shoot an elephant with a peashooter.[14] At length, however, the damaged reactor was partially smothered with 5,000 tonnes of material.

The resolution of one crisis led directly to another. The second concern was that underneath the damaged reactor lay a water-basin, a huge two-tiered structure designed for the reactor's emergency cooling system. After the explosion, this basin had reportedly been filled with contaminated water. After the stopper had been placed over the reactor, the authorities became aware that the 5,000-tonne sand covering threatened to push the 190-tonne white-hot reactor downward into the basin—or bubbler pond—reacting with the water to produce radioactive vapour. There seemed to be a real danger, in the view of Velikhov and others, that the core would then melt through the concrete layer under the basin and contaminate the ground water, i.e., it might have followed the course of the so-called China-Syndrome.

It was considered an "urgent requirement" therefore to ascertain how much water remained in the bubbler pond, and to determine its radioactivity and how to channel it away from the damaged reactor. Evidently, "hundreds of firefighting appliances" were employed for this purpose, but it proved impossible to remove all the water from the pool. A mission of equal danger to that carried out by the helicopter pilots ensued: to open two slide-valves that would enable the water to drain out of the pond, in pitch-black conditions in an area akin to a huge bathtub amidst very high levels of radiation.

The onus fell on three "volunteers." A. Ananenko, a 27-year-old senior engineer from the second reactor unit knew where the valves were located. He could open one and show another engineer, V. Bespalov, where the second valve was located. B. Baranov, a shift supervisor at the plant volunteered to hold the light. The initial attempt reportedly failed because the men, who took radiation monitors and carried lamps, could not reach the valves when the amount of water increased too rapidly along one of the corridors. Subsequently, the men's light failed and they were forced to grope in the darkness until "by a miracle," they found the valves, which turned instantly, releasing the water.[15]

If the major crisis had been averted, as Velikhov stated, the authorities were, to say the least, fully occupied with decontamination work. To cool down the reactor and to render the fourth generating unit as safe as possible, cranes, bulldozers, drills and trucks were required in huge quantities. Even while the bubbler pond was being emptied, for example, two convoys of trucks had left the Iaroslavl Auto-Transport Association carrying 700 tonnes of red lead for restoration work at the Chernobyl plant.[16] The means to avert a disaster were actually the cause of the most pressing problem that arose, namely the churning up of dust by vehicle wheels. "Every few minutes: trucks were bringing concrete to the damaged reactor area and armoured personnel carriers were also operating constantly."[17]

The same situation was to be found in other areas of the 30-kilometre zone. V.V. Vetchinin, Chairman of the Epidemiological Department of the Ukrainian Ministry of Health, remarked in late June that the hot weather had increased the amount of dust in the atmosphere, necessitating "increased health and sanitary measures." Contrary to popular opinion, he stated, laboratory tests had proved that rainwater was not dangerous, and that rainwater collected had not carried radioactive particles.[18] Because of the dust threat, signs were placed along the roadside all the way from the nuclear plant to the start of the "clean zone," warning drivers not to use the "hard shoulder" because they would churn up the dust that had collected at the verge.[19]

Vehicles leaving the disaster area had to undergo at least two "scrubbing downs," the first approximately six kilometres from the reactor, and the second at the zone border. Trucks coming from the "outside" unloaded their cargo at the border without entering. In the city of Chernobyl, the Voroshilovhrad section of the Ukrainian State Motor Vehicle Inspectorate supervised the hosing down of streets, and a chemical defence brigade was occupied with washing the leaves on the trees and the walls of apartment blocks.[20] Despite the use of unmanned vehicles, armoured personnel carriers and the regular washing down of every area, those involved in the clean-up gradually acquired high doses of radiation during their work.

Soviet Deputy Premier, Ivan Silaiev, noted that the clean-up workers gradually accumulated a dose of over 10 roentgens of radiation, at which point they were obliged to leave the disaster zone and not permitted to return to work. He declared that the "international annual norm" of 25 roentgens could be attained in the Chernobyl zone in two or three weeks.[21] In practice, these regulations were not always applied firmly.

This is clear from the following examples: on 6 June, Leonid Ilin, a member of the USSR Academy of Sciences, reiterated Silaiev's remark cited above that all workers were removed from the job when radiation levels reached a high level, but now he set that level at 25, rather than 10 roentgens.[22] On 17 May 1986, Izvestiia carried a detailed account about clean-up work, particularly at a trench located close to the damaged reactor, where drill operators were carrying out "very dangerous work." The correspondent said that the radiation monitors at the entrance to the showers indicated that everything was normal, but added the significant qualification, "normal for the special zone, that is." In short, then, at some point between 13 and 17 May, assuming the accuracy of the reports, maximum radiation doses were raised for people working in the danger zone.

On 5 June, a Ukrainian newspaper reported that although the fourth reactor had been corked at the top, "radiation is still escaping from it." It was therefore possible to approach this area "only in armoured vehicles." People were working there, but were in covered vehicles with an additional lead protection.[23] The report was then contradicted by the *Vremia* newscast on Soviet television only two days later. Correspondent V. Pisarevsky informed viewers that there was no longer any need for special posts to arrange the transportation of workers in the armoured cars to the nuclear plant, "one can drive up to the station's administration and service unit in an ordinary car."[24] It is possible that Pisarevsky was trying to assure his audience that the radiation situation was improving, but on the face of things, workers again seem to have been exposed to danger unnnecessarily.

Another example concerns a mishap that occurred a day or two after the explosion, when firemen were pumping water out of the basin beneath the reactor. Evidently a dump truck ran over a hose coupling and firemen were hit by a fountain of gushing contaminated water. The damaged hose was replaced but "the wet got through." The firemen were examined by doctors who found no major problems, but nonetheless ordered them home. The firemen evidently refused and were consequently allowed to continue working.[25] One can commend the bravado, but not the laxity of the doctors in allowing the injured to return to work. The nature of the clean-up work entailed a considerable amount of self-sacrifice, some of which—in view at least of the reportedly huge number of volunteers from as far away as Siberia[26]—could have been avoided.

Before the end of May, work was under way to decontaminate the first and second reactors of the Chernobyl station.[27] About fifty shift workers were said to be on duty at the first three units in mid-May, excluding repairmen. These units were shut down but were being closely monitored.[28] A number of nuclear engineering specialists were at the site, apparently in order to assess whether it would be possible to start up the first two reactors in the immediate future, a sign that energy questions remained uppermost in the minds of the authorities, even during a hazardous clean-up operation. The specialists included V.K. Bronnikov, the chief engineer at the Minsk nuclear power and heating plant; E.S. Saakov, chief engineer with an association of the Ministry of Power and Electrification of the USSR; T.G. Plokhy, formerly the deputy chief engineer at Chernobyl who later moved to the Balakovo nuclear power plant; and E.I. Ihnatenko, the deputy chief of the *Soiuzatomenergo* association. Silaiev foresaw the "rehabilitation of Chernobyl atomic power plant within a certain time,'' albeit without the fourth unit.[29]

The threat of dust was combatted by two principal methods. First, helicopter pilots dropped a liquid synthetic rubber over the dusty areas close to the fourth unit. The rubber evidently hardened when it came into contact with the air to form a ''reliable film'' over the dust. Second, in areas where radiation levels were somewhat lower, the decontamination units used an absorbent liquid, reportedly similar to that used in making polythene bags. The liquid was sprayed over the dusty areas, absorbing the dust and subsequently cooling into a ''film-like substance,'' which could be collected and taken to other areas for disposal.[30] Within the disaster zone, on 20 May, the soil was overturned and sprayed with a disinfectant by helicopters. Equipped with three tanks, the helicopters were reportedly capable of unleashing about 12,000 litres of liquid in five minutes. This was the first of many substances applied to the soil. The Soviets at this early stage seem to have decided that it might still be possible to maintain the land for agricultural use at some point in the future.[31]

By 25 May, Evgenii Velikhov, who was the chief scientist involved in the clean-up operation, said that the main task confronting workers in the zone was to ''prevent groundwater contamination.''[32] Construction workers at the fourth unit faced the immense task of intercepting underground waters south of the reactor to prevent them from being contaminated. The main goal was to prevent ''run-off water'' from reaching the Prypiat River, which flows into the Kiev Reservoir, the main water supplier to the city of Kiev before the accident. The authorities therefore began the construction of a 20-kilometre long embankment along the Prypiat River to protect the water's purity.[33]

Although the rainwater may have been declared safe by a Soviet scientist, the Soviet authorities, with justice, did not take this conclusion for granted. They put into operation long-practiced methods for controlling the environment, harnessed originally to try to ensure better weather for Soviet harvests (not with outstanding success, it must be admitted). Large AN-12 planes were used in the latter part of May to disperse the clouds around Chernobyl and prevent the rainfall, which, it was feared, not only might carry radioactive particles, but would wash contaminated soil into the Prypiat River. Special ''re-agents'' or powder-like mixtures were allegedly used for this purpose.[34] Evidently the measure was successful, but the dry weather, as shown, led also to an increase in the amount of dust.

The Chernobyl area, as part of the Polissian marshes, contains extensive areas of peat bogs. The has a high retention level of radioactive particles, which intensified the threat to the water supply. Because of the dry weather, however, a number of fires broke out in the peat bogs, which were said to be impeding work at the damaged reactor site in early June 1986.[35] The marshland area runs northward into Belorussian territory, and it was in the latter area that one of the major crises in the aftermath of the Chernobyl disaster occurred.

As noted in the previous chapter, the town of Bragin, in Gomel oblast, was not evacuated in the days after the accident, ostensibly because there was simply nowhere to house its residents, all the available places having been taken by evacuees from the areas bordering with Kiev oblast of the Ukrainian SSR. Bragin is located about 50 kilometres north of the Chernobyl nuclear power plant, and was thus well clear of the 30-kilometre danger zone. Belorussia, however, had been in the direct path of the radioactive cloud that was blown northwestward after the 26 April accident and evidently it was also affected by the wind in the days afterward.

The working group to "eliminate the accident's consequences" in Belorussia was headed by A. Petrov, Deputy Chairman of the Belorussian Council of Ministers. Early in June, the group discovered an area of high radiation outside the 30-kilometre zone. On the map of Gomel oblast that was being used to delineate dangerous regions, they were obliged to draw two lines: the first for the official 30-kilometre zone, and the second for the real danger zone, "after on-site [radiation] measurements had been taken." The population in the latter area was instructed not to consume food from the small private plots that each household was allowed to cultivate outside the collective and state farm system. Water supplies had to be hurriedly investigated: some wells were sealed, while others were cleaned out thoroughly. New artesian wells were sunk, public roads were asphalted and contiguous territory was covered with plastic film. Agricultural workers were forced to wear breathing apparatus and hermetically sealed cabins were quickly prepared for tractors.[36]

As noted above, the situation was serious enough in parts of Gomel oblast to merit regular visits from the Belorussian First Party Secretary, Sliunkov. For a period of time—it may been either one week or up to one month—the population had been exposed to substantially higher levels of radiation than hitherto believed. And the most serious threat was to the 7,000 residents of Bragin who had not been evacuated at the end of April. In June, the Belorussian authorities decided belatedly to decontaminate the town without removing its citizens. The situation was made worst by the preponderance in the area of peat and sandy soils known for their high level of retention of radioactive particles.

The entire town had to be scoured from top to bottom. Approximately 1,220 houses were cleansed, i.e., they were found to have been contaminated. Thousands of trees were felled (the leaves were irradiated), sheds, garages and storehouses were cleaned or in some cases destroyed. All 169 wells in the town had to be cleaned, and all food was brought in from the outside. Radioactive particles had fallen into streams, which were irradiating garbage dumps and cow manure. The military were called in to help resolve a desperate situation. The local population appears at times to have been quite oblivious of the danger, allowing cattle to feed in ditches and cutting hay in proximity to signs warning people to keep clear of the verges.[37]

Late in May 1986, the Deputy Chief of the Sanitary-Epidemiological Department of the Belorussian Ministry of Health, V.N. Buriak, was interviewed in *Selskaia gazeta*. He reported that there had been a noticeable increase in radiation levels in Bragin, Khoininki and Narovlian raions, but that subsequently it had declined to a harmless level. Farm work was therefore continuing, with workers using respirators and changing their clothes constantly. As for dairy products and green vegetables, Buriak's attitude was casual. Products made from milk were said to be completely safe, but more care was required with eggs. Green vegetables sold in stores were harmless, but home-grown vegetables should be pickled. Within three weeks, said Buriak, all vegetables grown near the Chernobyl station would be safe to eat. The remark can be compared with other official comments that nothing should be eaten from the fields around the reactor. Water, on the other hand, was unsafe in many areas, according to the Belorussian official, because local residents had not taken the precaution of sealing wells. But had they known about the necessity of such a step beforehand?[38]

An important proviso should be made concerning the Belorussian situation, which appears from reports to have been worse than in Ukrainian. Such an assumption is unwarranted. The "hotspots" of high radiation occurred also in Ukrainian territories outside the 30-kilometre zone. There is evidence, for example, that a group of Latvians assisting in the clean-up at Chernobyl were ordered to take precautionary measures such as hosing down the camp every morning, even though the official danger zone was 20 kilometres away.[39] The difference between the two Soviet republics was essentially in the reporting. One can only speculate about the reasons behind this, but reports from or about Belorussia were more open and frank than those from Ukraine, and especially in the local Ukrainian newspapers. The latter were reserved and usually carried optimistic messages throughout the text testifying to the constant improvement in the situation and the "significant" daily lowering of the radiation levels. At the extreme, comments bordering on the ludicrous resulted from this combination of warning and false optimism. Take, for example, some of the statements made to a Ukrainian newspaper by the Chairman of the Epidemiological Department of the republican Ministry of Health:

The radiation level has fallen significantly...but *increased* health and sanitation measures are being taken....Almost all garden produce may be consumed without fear....Gardeners should avoid making dust, eat indoors and wash their hands before meals. There is no problem with swimming in the Dnieper, but if you go to the beach, out something underneath you. Don't lie on the bare sand.[40]

Because of the nature of this reporting, Western observers, and undoubtedly Soviet citizens, were often puzzled by the turn of events. The above official had emphasized that one could swim in the Dnieper, but before long, the Soviet authorities declared that "the danger has risen of radioactive particles getting into the Dnieper,"[41] thereby acknowledging that the Dnieper was hardly a suitable area for swimming, as the health official had stated. Not until July however, was a new pumping station constructed to carry water to Kiev from the Desna River, a "reserve water system."[42] The Dnieper River, which supplies 80 per cent of the water supply for the Ukrainian SSR, was no longer considered safe enough to supply the citizens of Kiev. But had it been safe initially and subsequently become contaminated? Or had it been affected initially by the radiation cloud?

Special equipment was required for decontamination work, and much of it had to be manufactured anew or transported from various parts of the Soviet Union. Soviet reports that all areas of the Soviet Union assisted those people at Chernobyl, which began to appear on a daily basis in most newspapers after the first week of May and throughout June, were accurate. Outside assistance, quite simply, was a necessity. If one considers how much equipment arrived in the weeks after the disaster and in the month of June, then it is clear that the task force encountered the worst hazards in the first days, when radiation was at its highest and very little equipment was available.

On 19 May, it was reported that a large remote-controlled bulldozer had begun to operate in the clean-up work at the nuclear plant. It was equipped with radio controls to replace the human operator in order to clean away the debris "in areas where radiation levels are still too high for humans." It could be controlled from a car standing at a distance of 150 metres. A second bulldozer was said to be in preparation.[43] On 4 June, the Izhmash production association evidently sent a remote-controlled transport robot to Chernobyl that had been built in five days and was intended to work directly next to the disabled reactor.[44] Other equipment destined for Chernobyl included 70 mobile four-kilowatt electric generators from Kaluga, telephone exchanges from Sverdlovsk "to enable the completion of telephone links at settlements adjacent to the

AES," and mobile homes for the workers, manufactured in Novgorod. [45] The mobilizing of all resources to make work near the fourth unit safer illustrated the magnitude of the ordeal: short shifts in areas of high radioactivity at an incalculable cost to future health of the workers involved. In early June, for example, a 1.4 kilometre road was completed to the fourth reactor, presumably to facilitate transportation of materials to the site. The maximum length of time that could be spent working on it was said to be two hours. [46]

By the second week of May, the Soviet authorities had decided to entomb the damaged reactor inside a concrete shell or "oesophagus." On 13 May, Ivan Emelianov, the Deputy Director of the Institute for Energy Technology, informed foreign reporters in Moscow that the reactor would have to be buried in concrete for hundreds of years, until all the radioactive elements had decayed to a harmless level. [47] Four days later, Ivan Silaiev appeared on Soviet television, and informed viewers that the temperature of the reactor had now cooled down to 200–250 degrees celsius. The main task ahead, he stated, was to create a sarcophagus, a huge container that would enable the task force to bury all the radioactive fallout. [48]

Initially, a tunnel was to be built of the sort used in underground railways, using the assistance of Soviet coal miners. The Minister of the Coal Industry of the USSR, M.I. Shchadov, arrived at the accident site, along with the Ukrainian Coal Minister N. Surgai and a team of 388 miners, made up of 234 from the Donetsk coal basin and 154 from the Moscow Basin. [49] Their goal was to construct a tunnel of 160 metres from the third reactor to the damaged reactor, in order to install a 30-metre square concrete slab underneath the damaged reactor. Shchadov explained that not only was this a "huge and complicated task" in itself, but that matters would be made worse because of the waterlogged sand that the miners would have to dig through. Shchadov's prognostications resembled a Soviet plan commitment. The estimated time we will need, he remarked on *Radio Moscow* is 45 days, but because of the urgency we have been assigned only 30 days. He and the miners, however, had pledged to complete the task in only 25 days. [50]

To assist the coal miners, 200 Kiev metro-builders were dispatched to the site, to be followed by other workers considered by the authorities to have relevant experience: Moscow metro-builders, drilling and tunneling specialists, surveyors from Kharkiv, Gorky, Kuibyshev and Dnipropetrovsk, and workers from the Baikal-Amur railroad. [51] While the miners drilled their way slowly toward the reactor, other workers attempted to control the temperature of the area below the concrete shield under the reactor, "to ensure reliable supervision of the miners' work below." [52] Once the miners had completed their tunnel, an "extra cushion" was to be built under the shield with a "powerful cooling system" fitted directly into the reinforced concrete.

In Soviet reports, the task was compared to traditional work at Soviet coal mines, at which, it was said, work was also hazardous.[53] In reality, the difficult conditions at Soviet underground coal mines notwithstanding, there was no comparison between the two. At Chernobyl, the miners worked in three-hour shifts, shuffling forward at a rate of some sixty centimetres an hour. The rates were measured less in centimetres, however, than in terms of the reinforcement rings that were put in place to line the sand along the underground passage. Each ring took about 15 minutes to put in place.[54]

Toward the end of the tunnel, the miners suddenly began to cover greater distances. On 19 May, *Radio Moscow* announced that the current rate of 14–15 metres of tunnelling each day was double the record attained underground under "normal conditions." One newspaper stated almost a week later, however, that "dozens of metres" remained to the fourth reactor,[55] implying that the coal miners might not attain their scheduled progress. Yet two days later, the tunnel was completed, well ahead of schedule.[56] It had taken only 10 of the 25 days anticipated by Shchadov. What had occurred in the interim? Had reinforcements arrived? The questions are important in view of the fact that workers involved were incurring radiation doses of up to 3–4 roentgens per hour. If one applies the maximum standard of 10 roentgens, which was used initially for workers at the accident site, then even the allotted three hours may have exceeded the maximum norms. Yet even when records were being set earlier in the work, it had still fallen behind schedule. So how it was completed so quickly?

One reason may have been that the authorities exceeded the maximum working norms for a zone of high radiation. The miners, it is reported, worked in eight shifts of three hours a day. V.A. Brezhnev, the Minister of Transport Construction of the USSR, who was at the site organizing the transport, said that at the trench marking the tunnel entrance (at which the drilling began), the maximum time limit for work was five hours. The ten-minute trip from his headquarters to the trench (which was by the third reactor, it should be recalled, 160 metres from the damaged reactor), "was fraught with danger."[57] The drilling operators, then, according to the information of a Soviet official, were incurring double the maximum daily dose of radiation, even if one assumes that each shift worked only the one day and was subsequently replaced.

Work in the administration building may have been safer, but being also relatively close to the fourth unit, it was not as safe as being in Chernobyl itself, or Bragin, for example. Yet here, the working time was extended to twelve hours, during which, according to a duty office working for the Ministry of Transport Construction, "we sometimes have to go to the [accident] site too."[58] In short, then, the time limits imposed for working in hazardous conditions were not imposed rigidly. The main goal was to reduce the danger that existed from the reactor; essentially the safety of those working at the site was a sedondary concern. It can be argued that the safety of the Soviet population as a whole was at stake, but this should not detract from the highly dangerous and excessively long hours of work being accumulated by the task force workers.

Soviet reports have maintained that none of the "tomb workers" have suffered any ill effects from their work, a comment that is surely premature. There is evidence to suggest that many of the workers themselves were neither volunteers, nor happy with their situation. The Latvian newspaper *Cina*, for example, noted that the workers assigned to Chernobyl were simply selected from lists of military reservists, and experienced apprehension as they were transported to Ukraine. Their camp was said to be a grim place, in which "nobody smiles."[59] Some of the miners—and the implication is that they were from the Moscow Basin—had been "expelled" from the area for anti-social behaviour, most likely for alcohol consumption, which was not permitted at Chernobyl.[60] This can also be taken as a sign of the tension under which the miners were working.

Early in June, the main underground tasks were completed. Pipes were fitted to the tunnel, along which cement could be poured under and around the fourth reactor. Army sappers were used to blast openings on three walls of the ventilation shafts, having carried out experimental blasts on concrete walls "at a safe place," to ensure that the ruined reactor set would not be shaken.[61] Liquid nitrogen had previously been pumped underneath the reactor to reduce its temperature further. After 1 June, the major task was to continue constructing the sarcophagus below the reactor, entombing it for posterity as a high-level waste, and a monument to the Chernobyl disaster.

The main drama of the Chernobyl disaster had ended. Over a period of five weeks, the chances of a much worse disaster had been eliminated. Decontamination work continued, however, and thousands of people were uprooted from their homes with little prospect of an early return. The political and economic repercussions of the Chernobyl accident were only now beginning to be felt.

The Political Consequences of Chernobyl

There are two political aspects of the Chernobyl disaster that should be considered. One concerns what might be termed a high-level political dispute between Moscow and Ukrainian officials. The second, which is more clear-cut, encompassses dismissals as a result of the accident. If the accident was to be blamed mainly on human error, as early Soviet accounts suggested, then some dismissals were inevitable. On the other hand, the fault could have been laid on technology or defective equipment, which had been a problem in the past. To take this course, however, would have jeopardized the industry's future, since it would have implied that other Soviet plants were equally dangerous.

It is clear that there were some differences between regional and Moscow officials in the first days after the accident. E.I. Ihnatenko, the Deputy Chief of the all-Union *Soiuzatomenergo*, stated that when he flew to the accident site from Moscow on 26 April, "he had never realized that the situation was so serious."[62] Valentin Falin, the Director of *Novosti* news agency, informed the West German magazine *Der Spiegel* that Gorbachev had not received a detailed report on the Chernobyl disaster until two days after the accident. The first reports given to Moscow, Falin maintained, were incomplete and "turned out to be incorrect."[63]

Further, after Ligachev and Ryzhkov had visited the site on 2 May, additional measures were taken by the CC CPSU Politburo, implying that those that had been adopted at the local level previously were inadequate.[64] In Chapter One, it was shown that Liashko appeared to resist this line of argument by stating that initially there did not seem to be great cause for alarm. In mid-May, *Pravda* also joined in the discussion, criticizing the delays in the releasing of information about the disaster, "which contributed to people's concern." The main lesson to be learned, it concluded, "was that it is necessary to trust people."[65] The complaints about the paucity of information were again directed ostensibly at local officials.

And yet, various circles in Moscow were aware from the first of the import of the accident. At a press conference in Moscow on 19 May, Emelianov stated that the Government Commission appointed by the Soviet government began its work on the day of the accident, i.e., 26 April, a full two days before any announcement was made about the accident.[66] This implies that the delay in reporting emanated from Moscow rather than Ukraine. Also, health officials were dispatched to Poland, helicopter teams were ordered to fly to Prypiat, and the head of Moscow's Hospital No. 6, Guskova, was notified on the same day as the accident and sent out a team of medical workers. Viktor Sydorenko, the First Deputy Chairman of the USSR Committee for Safety in the Nuclear Industry, also said that the information "with requests for an emergency team" was received in the early hours of the Saturday morning, after which senior officials of the Committee had at once left for Chernobyl.[67]

These factors all point to one conclusion: the Moscow authorities knew that a major catastrophe had occurred, but chose to lay the blame on regional authorities for the delays in reporting the accident and in taking adequate precautions. As far as the USSR's international standing was concerned, this was an expedient manoeuvre. The Moscow authorities could claim that they were not fully aware of the scale of the accident at Chernobyl. The policy had some obvious flaws, however. As a Western analyst wrote, it was "stretching the point a bit to suggest that Kiev could not or would not reach Moscow by telephone."[68]

Some Western writers felt that the logical reasoning behind Moscow's policy was to use Chernobyl as an excuse to conduct a purge of the Ukrainian party leaders. They wrote that the Ukrainian First Party Secretary's job was on the line, that Shcherbytsky, the last relic of the Brezhnev years, who had somehow survived the Ukrainian Party Congress of March 1986, would now be made the scapegoat for Chernobyl. This deduction was based on some dubious assumptions about the Ukrainian party chief's alienation from the Gorbachev regime, and about the way the Soviet system works.[69] Whether or not Shcherbytsky is eventually removed, and whether or not that removal is related to the events surrounding Chernobyl, there is little concrete evidence that he has ever been in disfavour since Gorbachev became General Secretary, or that his fall is inevitable in the long-term because he was reputedly an ally of the late Brezhnev.

Following his visit to the damaged fourth reactor site with Ligachev, Ryzhkov and Revenko, Shcherbytsky fuelled speculation about his position simply by his absence from the danger zone in subsequent days and his infrequent comments about Chernobyl, which he left largely to Revenko. Although Shcherbytsky had visited evacuees rehoused in Borodiansky and Makariv raions, according to a report of 20 May,[70] little appeared about him in the press for the next six weeks. Then on 1 July, he and the Ukrainian premier, Liashko, visited the plant site and several areas of the Chernobyl, Poliske, Ivankiv and Borodiansky raions that had taken in evacuees. Revenko and the head of the Kiev oblast executive committee, I.S. Pliushch, also took part.[71] In Chernobyl, Shcherbytsky met with V.K. Gusev, who had replaced Borys Shcherbyna as the head of the government commission investigating the accident.[72] On 11 July, a Plenum of the Central Committee of the Communist Party of Ukraine was convened in Kiev to discuss Chernobyl and specifically its impact on energy and other sectors of the economy. Shcherbytsky was the main speaker.[73] Evidently, he had either survived the disaster or, as seems more probable, he had never been in any serious danger of losing his position in the first place.

At the same time, the authorities in Moscow clearly had given themselves the scope to carry out dismissals among the Ukrainian party hierarchy. They emphasized on various occasions that the Ukrainian party personnel had misjudged the situation, and that many of the early problems might have been eliminated had the proper steps been taken from the first. The timing of dismissals in the USSR is always imprecise, however. The case can be cited of the Ukrainian steel minister, Dmytrii Halkin, who has been hounded for an inept performance in the republican steel industry for over two years, receiving reprimands in the process. He has, however, retained his position in the face of disastrous output returns, and was even awarded the Order of the October Revolution on the occasion of his sixtieth birthday early in 1986.[74] Removal of officials in the USSR rarely follows a logical pattern and making predictions is therefore often risky.

One can posit that the likely target for the Moscow authorities was the Kiev oblast First Party Secretary, Hryhorii Revenko, but he had been appointed to his post only in November 1985,[75] and the Soviet authorities may have been unwilling to replace a recent appointee. Only at the lowest levels of the party structure—at Prypiat and areas of Chernobyl raion—were dismissals made. The impact upon the Ukrainian party in the first months of the accident was therefore negligible. Prominent party leaders were thrust "into the firing line," but no shots were fired.

For local plant officials, however, there was no leniency. On 12 May, *Pravda* announced the dismissals of three party members who worked at the nuclear plant: O. Shapoval, O. Sichkarenko and O. Hubskii. A few days later, it was revealed that Halyna Lupyi, the head of a Komsomol organization at the station, had "run away" when the accident happened and could not be located for nine days. Evidently she returned only after a telegram had been sent to the home of her parents. The deputy head of the youth section of a construction division, Iurii Zahalsky, had also reportedly "shirked his duties" and spent the post-accident days "attending to his personal affairs." Zahalsky was fired from his job.[76]

It was revealed in subsequent comments that some of the plant officials had not merely failed a unique test, which was probably correct, but should not even have been allowed to become members of the Communist Party in the first place. Revenko was quoted in the weekly *Ogoniek*:

> A sharp examination of each individual is being carried out. We have already got rid of a few people, including people in leadership positions. They have parted with their party tickets, these people got into the party by chance, and they couldn't even withstand the first test.[77]

How had these officials failed their test? In the first place, many had panicked. The "shirking of duties" that was cited in the cases of Shapoval and Zahalsky may have implied simply fleeing from the scene, a sign that the problems of discipline in the local party organization discussed in *Pravda* during the previous summer had not been eliminated. There are several indications that many party members abandoned the area as soon as the scale of the disaster became known. *Sovetskaia kultura* noted in late May that one party member at the plant had fled to Odessa after the accident. Slava Staroshchuk sent his fellow workers a telegram asking them to send the money that they owed him.[78]

When the Prypiat urban party committee moved back to Chernobyl from Poliske in June, it reported that 177 party members had gone missing.[79] Two weeks later, *Pravda* stated that "Because of defects in organization and training, to this day a portion of the workers is to be found on the run. Indeed, among them are shift supervisors and senior operators."[80] The extent of the panic was hinted at in a comment by *Novosti* in late May 1986 that "There was a danger of scaring people into panic which could result [sic] from misinterpreted excessive information."[81] Emelianov also told an interviewer from the Italian Communist Party newspaper *Unita* that the authorities had decided to release a "selective form of information" about the accident in order to forestall outbreaks of panic.[82]

Although the Ukrainian party did not suffer greatly as a result of the Chernobyl accident, the same cannot be said of various ministries, or those officials at the Chernobyl fourth unit who were occupying positions of responsibility at the time the explosion occurred. The initial target was the USSR Ministry of Power and Electrification, which not only was responsible for the Soviet nuclear power industry as part of the overall energy sphere, but also was assigned numerous tasks in the clean-up operation, such as receiving and accommodating those arriving in the zone, and organizing catering, consumer and medical services for them.

The ministry was held responsible for the chaotic situation that emerged after the accident as well as for the event itself. The party leaders of Prypiat and Chernobyl raion soon began to complain about ministry officials who were reportedly doing little about decontaminating the town of Chernobyl. Plans for the work, it was reported, were at the raion executive committee's headquarters. The ministry had been informed of this, but had not responded. According to N. Stepanenko, the Deputy Chairman of the Kiev oblast executive committee, the ministry's officials were essentially "passing the buck," and spending an inordinate amount of time in discussions, "during which selfish attitudes manifest themselves." The general feeling among local party officials was that the Ministry of Power and Electrification of the USSR—rather than its Ukrainian branch, it should be emphasized—showed a lack of interest in people's housing, domestic living conditions and food.[83]

It soon became clear the the Ministry of Power and Electrification was also to pay a price for the disaster itself, although it was easier for the authorities to apportion blame for failings in an area that virtually ensured problems, namely supervision of the clean-up operation. On 20 July 1986, the Politburo of the CC CPSU issued a statement about the Chernobyl accident, which announced the dismissal of the First Deputy Minister, G. Shasharin, and gave a severe warning to the Power Minister, A. Maiorets, who "deserved" to be relieved of his duties in view of the serious defects permitted in the supervision over the nuclear plant, but was being let off with a severe reprimand in view of his short-term of office (one year).[84]

Along with the State Committee for Safety in the Nuclear Power Industry, the Ministry of Power and Electrification was declared by the Politburo to have been guilty of gross negligence and "a lack of control" over the nuclear plant. Neither had taken proper steps, it was alleged, to comply with safety regulations, or to stop the breaches of discipline and operating rules. The nuclear power industry was assigned to a newly established Ministry of Atomic Power Engineering, the report continued. Evgenii Kulov, who had been appointed head of the newly formed Safety Committee only in August 1983 (see Chapter Five), also lost his job on 20 July, along with a Deputy Minister of Medium Machine-Building, and Emelianov, who had been one of the main Soviet spokesmen commenting on the disaster almost up to the day of his removal from office.

Along with officials of ministries, the leaders of the Chernobyl plant were severely disciplined and, moreover, held largely responsible for the apparent underestimation of the extent of the accident at the outset. At a meeting of Kiev oblast party committee in June, the station's Director, V. Briukhanov, and Chief Engineer, N. Fomin, were dismissed for their failure "to assess correctly what had occurred and to take adequate measures to organize efficient work in all areas to eliminate the consequences of the accident." They had exhibited "irresponsibility and mismanagement," declared *Pravda*.[85]

Blame was also apportioned on other plant leaders. One Deputy Director, R. Soloviev, had abandoned his post "at the most difficult moment," while two other Deputy Directors, I. Tsarenko and V. Hundar, had also reportedly failed to carry out their duties. The work of the station's trade union was also declared to have been inadequate. E. Pozdyshev was appointed as the new Director of the Chernobyl plant,[86] but the other positions were not filled at that time. Subsequently, both Briukhanov (20 July) and Fomin (31 July) were also expelled from the party.[87]

Following the extensive criticisms of the Chernobyl nuclear power plant in the Soviet press, it cannot be denied that there were some serious management problems there. At the same time, officials such as Fomin had been allowed to make various claims about the safety of the station on *Radio Kiev*, in *Soviet Life* and elsewhere. Fomin and Briukhanov were in clearly not in danger of losing their jobs before the disaster.

Can one say that they were scapegoats? Certainly one can say that many of the problems that had arisen at Chernobyl were a result of party policy rather than of petty officials. The rapid expansion of the Chernobyl plant was pushed through, for example, without adequate quality control in accordance with CPSU's decision to expand the nuclear power industry, with Ukraine as the key area, and Chernobyl as the republic's largest station. The dilemmas at Chernobyl, then, cannot be divorced from party politics. Yet Soviet statements carefully avoided any criticism of party policy in the nuclear industry. The accident also was not an event that occurred solely because of the incompetence of the plant's officials. The main question therefore is whether the disaster was a direct or indirect consequence of the Soviet nuclear energy policy.

The Future

The long-term impact of the Chernobyl disaster on Soviet agriculture is not likely to be substantial in terms of the overall Ukrainian harvest, as the raion is not a major grain-growing region of Ukraine. But given the network of rivers that connect the zone to grain-growing regions, a key concern was radioactive pollution of these rivers. A virtual "dead zone" was anticipated by some Western observers. Evgenii Velikhov had stated frankly during an interview on *NBC*'s "Meet the Press" programme that agricultural cultivation within the 30-kilometre zone was out of the question.[88] Many in the West drew the same conclusion: the Chernobyl raion would barely affect Ukraine's agricultural output, but the land could not be used for farming in the future, were the area to be repopulated.

Generally, the Soviet authorities did not share Velikhov's assessment. Emphasis was placed on the "normalization" of the 30-kilometre zone and those villages bordering on this zone. Indeed the local party officials did not delay the return of evacuees to their homes as soon as it was deemed possible. Areas were reportedly totally decontaminated and then resettled, even while in areas further to the north, later evacuations occurred. In brief, the Soviets made a hasty decision in some cases to return families to their homes and even to resume farming, at the beginning of June, barely five weeks after the disaster.

For example, on 3 June, A. Shchekin, the chairman of the Chernobyl raion executive committee, announced that as the radiation level had fallen in several areas of the raion, it had been decided to send some evacuees back to the villages of Zamoshnia, Bychky and Hlynka. These consisted of 260 families that had been employed at a state farm and a collective farm, but rehoused temporarily in Borodiansky raion. A depot was being established at a nearby collective farm to prepare equipment "which will be used to plough the land."[89] The three villages are all located in the southwestern part of Chernobyl raion on the very fringe of the 30-kilometre zone, but this hardly detracts from the surprising announcement so soon after the statement of a prominent scientist like Velikhov.

Other Soviet officials expressed optimism that the region had an agricultural future. Two days after the statement concerning the return of the evacuees, Iu. A. Izrael appeared at a Moscow conference, and responded confidently to a question from a Japanese newspaper: are there any zones that have become unfit for human life for a long-term period or even for ever?

> This is a very serious question...we are studying the situation, especially the isotopic composition. I can say that an absolute majority of the contaminated area will be restored to the economy, people will return there. But it is possible that some small area will be subject to further study.[90]

No future "dead zone" was foreseen, therefore.

On 19 June, *TASS* announced that farming had resumed on six farms near the damaged reactor, after experts had first checked the fields. Potatoes and fodder were to be grown, and the farmers were to be rotated at ten-day periods. In some areas, layers of topsoil had been removed to reduce the danger of radiation. Early in July, *Radio Moscow* stated that evacuees had returned to the villages of Nivetske and Cheremoshna, in the eastern part of Poliske raion, just within the 30-kilometre zone. The returning villagers "immediately got on with farming work" at the large collective farm *Svitanok*, which had 10,000 head of cattle.[91]

Finally in Belorussia, seven villages were decontaminated and declared suitable for human habitation by 9 July.[92] One could gain an impression from the above statements that it is simply a question of time before the entire region is reinhabited and all the evacuees returned. Yet as at other stages of the Chernobyl episode, the Soviet reports have encouraged false hopes about the overall situation. The above villages were neither representative of the danger zone, nor carrying out farming according to what one can term "normal practices." Because of the continuing hot weather in the region, ploughing was a highly dangerous undertaking. And how could a crop that had been contaminated be harvested?

Plainly, the vast majority of evacuees have little prospect of returning to their homes in the near future, and those that have been sent home face a very uncertain future. Hermetically sealed tractors were being prepared for use on farms within the zone during July 1986, but, as *Sovetskaia Rossiia* pointed out, the wheat in the fields around the fourth reactor was contaminated and unfit for use, and the crops in the 30-kilometre zone had been irradiated and could not be harvested. Scientists had to resolve the problem of the removal of the longer-living radioactive elements from the soil.[93]

Shortly after the evacuation, the authorities organized the preparation of winter homes for the evacuees, which suggests a more realistic approach to the future. Builders from various parts of Ukraine were given the task of constructing 7,250 new homes in several villages of Kiev oblast by October, in addition to extending about 6,000 of the houses that were accommodating evacuees. The first 150 of these homes were ready for habitation near the village of Lodvynyvka, Makariv raion, by 2 August.[94] A virtually "new village" was built for evacuees at the southern end of the village of Nebrat, Borodiansky raion.[95] According to *Izvestiia*, every family evacuated from the 30-kilometre zone was to receive a separate house or apartment before winter, and 4,000 houses were to be built in the northern part of the Gomel region by 1 October. The reason behind this plan was said to be economic: the northern part of Gomel oblast was an area of labour shortage.[96]

In the Prypiat area, tourist boats were to be used to house those workers manning the first and second units in the autumn of 1986. The boats were taken off their regular routes and brought up the Prypiat River via the Dnieper River. Between May and August, workers in the vicinity had been placed either in the Lesnaia Polnaia sanatorium, which was some distance from Prypiat, or the Kazkoyi young pioneer camp, which was reportedly overflowing with up to three times as many personnel than it had been designed for.[97]

A more permanent abode for construction and nuclear power plant workers was needed urgently, however. Using the resources of the entire republic, a new town for 10,000 residents was being built on the bank of the Kiev Reservoir in July 1986. Called Zelenyi Mys (Green Cape), it was to be built as a miniature Prypiat, but consist entirely of men. Their families were to be housed in Kiev or Chernihiv, while the menfolk were to work 10–15-day shifts at the Chernobyl nuclear plant, and return to their families at intervals. In August 1986, a number of prominent officials toured the site of the new town, including Secretary of the Communist Party of Ukraine, B.V. Kachura, on 7 August; and Soviet Prime Minister, N.I. Ryzhkov and Chairman of the Committee for State Security of the USSR (KGB), V.M. Chebrikov, on 8–9 August.[98] Thus the nuclear plant zone was to be reinhabited, even if Prypiat itself was to remain a ghost town.

But the Soviets faced a seemingly impossible task to try to bring a defunct area back into agricultural production, and moreover, a region that was hardly essential to Ukrainian agriculture. Other than giving Soviet citizens an impression that some sort of normality had returned to the zone, one can only hazard guesses as to why they would move evacuees back into the dangerous zone, where irradiated crops lay unharvested, and where the future of the region was by no means clear. Possibly the overcrowding in temporary homes had made life unbearable for the evacuees and their hosts.

One thing is clear, however. The Chernobyl nuclear power plant was to be reactivated without delay, with the exception of the buried fourth reactor. If the area had been declared a "dead zone," then the future of the giant nuclear plant would have been in grave doubt. The importance of Chernobyl to the republic's nuclear industry can hardly be overemphasized. At the time of the accident about 45 per cent of Ukraine's nuclear-generated electricity stemmed from the station. Even without the fourth reactor, the fifth and sixth units were close to completion at the time of the disaster. It was not surprising therefore that statements were made by the party secretary in Prypiat, Oleksander Domaniuk, and others less than six weeks after the catastrophe that the first two Chernobyl reactors would be back in operation by October.[99] As far as the nuclear programme was concerned, Chernobyl had evidently changed nothing.

The impact of the disaster on the production of electricity was more significant than that on agriculture, in the short term. All the Chernobyl reactors had been shut down. Once the plant had been reactivated, however, the impact of losing the fourth reactor was negligible. To the Soviets, the more important concern was that Chernobyl might imperil an ambitious development programme. In mid-May, the Hungarian authorities stated that as a result of the accident, the amount of electric power supplied to Hungary by the USSR had declined by 1.5 per cent for the second week of the month, but that by the end of the year, the USSR would catch up on the shortfall.[100] Implicit in the statement is the premise that one of the new reactors coming on-line in Ukraine in 1986 could make up this deficit—most likely the Rovno plant's third reactor (autumn 1986) or the Zaporizhzhia station's third reactor (late 1986). As for domestic consumption, the large Trypilska hydroelectric power station near Kiev evidently compensated for much of the loss of power brought about by the complete shutdown of the Chernobyl station.[101]

After the disaster, prominent Soviet officials asserted their faith in the future of the Soviet nuclear industry. It was as though Chernobyl had released a surge of support for the industry in the face of mounting domestic and East European criticism (Western comments were less important in their eyes). Moreover, official statements changed noticeably in tone in May and June 1986, from cautious support for the industry to a more confident attitude, and finally, to a swaggering bravado that was quite astonishing in view of what had happened at Chernobyl. In late May, Viktor Sydorenko of the State Nuclear Safety Committee adopted the guarded approach:

> The accident will affect the nuclear industry substantially in some areas. But Soviet experts believe that the Chernobyl accident cannot and should not change the strategy of the nuclear industry. Proper lessons will be drawn, however bitter they might prove, and we will continue to advance. Progress cannot be stopped.[102]

One week later, V.A. Legasov, a Deputy Director of the Kurchatov Atomic Energy Institute of the Academy of Sciences of the USSR and a member of the presidium of the academy, was informed by a *Pravda* correspondent that letters opposing nuclear energy had been received by the newspaper. He defended an industry that was, he felt, essential to the future of civilization:

> I am profoundly convinced that atomic energy stations are the pinnacle of achievement of power generation....Nuclear energy sources are the beginning of a new phase in the development of civilization. They are not only economically advantageous compared to thermal stations in normal use and not only cleaner ecologically, but are the preparatory basis for the next technological leap. The future of civilization is inconceivable without the peaceful utilization of nuclear power. The likelihood of accidents is less than in unsophisticated systems, but if something does happen, the consequences are on a larger scale and harder to eliminate....But an accident thought unlikely did happen, and we must learn a technical, organizational and psychological lesson. People have been killed, but I am convinced that nuclear power will come out of this test even more reliable.[103]

Legasov's statement would have had serious repercussions had it been made in Western European countries, which were having doubts about the future of domestic nuclear industries after Chernobyl. It is unlikely that such a forthright statement, exhibiting unassuaged faith in the future of an industry after the worst disaster in its history could even have been attempted in most East European countries. Yet Legasov was assuring Soviet citizens, through their major newspaper, that not only would the industry continue in the future, but that it was the only route for the future, the one area from which a technological leap could be made. The day after his comments appeared in *Pravda*, E. Chazov reported that two more victims of the accident had died, bringing the death toll to 25, while Iliin said that 30 victims remained in critical condition. But economic issues were uppermost in the minds of the authorities.

Was there an alternative for the Soviet authorities? There was a body of opinion in the USSR that supported a revitalized coal or oil industry and the curtailing or reduction of emphasis on nuclear power at some point in the future. A.M. Petrosiants, the Chairman of the USSR State Committee for the Utilization of Atomic Energy was asked a similar question at a Moscow press conference. He maintained that the twentieth century was the era of exhaustion or approaching exhaustion of organic types of fuel. Of oil, gas and coal, only the latter had a long future, but 200,000 waggons of coal would be needed simply to match the output in kilowatt hours of the Leningrad nuclear power plant in a single year.[104] Coal, then, was feasible but impractical, although it has been noted earlier that the coal industry has its share of problems.

On 18 June 1986, Soviet Premier Nikolai Ryzhkov presented the Twelfth Five-Year Plan for 1986–90 to the USSR Supreme Soviet for approval. The output foreseen for nuclear power, at 390 billion kilowatt hours by 1990, had not been reduced from the target foreseen by the Twenty-Seventh Party Congress in March 1986. Moreover, Ryzhkov's speech gave no hint that plans had been affected by the accident at Chernobyl, and in fact he stressed that the "growth of atomic energy" would play a larger role in fulfilling the need for electric power in the years ahead.[105] In fact, increased production of electricity is essential to Gorbachev's plans for the "scientific-technological acceleration" of the Soviet economy that has formed the basis for development in the years ahead.

Evidently, the Soviet leaders can perceive no alternative to nuclear power. This in itself should not draw criticism. The safety record of the industry throughout the world, even after Chernobyl, is superior to that of its rivals in the organic fuels. What is questionable in the Soviet case is the apparent reluctance to raise questions in any depth about the plans for the future: its geographical development; the various problems that have occurred at most Soviet nuclear power plants in the recent past; the lack of qualification of many workers at nuclear sites; and the way in which the industry is organized (the new ministry notwithstanding). Above all, there was no attempt to reduce or even discuss the plans for nuclear power's rapid future development in the face of a major catastrophe.

Only in the area of safety have certain concessions been made to the International Atomic Energy Agency. Here, however, the amount of supervisory power that is to be allotted to the IAEA remains unclear. Moreover, since the traditional role of the IAEA has been to monitor nuclear plants to ensure that they are not being used for the manufacture of nuclear weapons, the involvement of the UN body was not unwelcome to the Soviets. The entire issue was linked with Mikhail Gorbachev's declaration that the Super-Powers should remove all nuclear weapons from the earth by the year 2000. Questions about safety and about the political future have been enjoined. There may be genuine reasons why the Soviet Union has begun to co-operate more closely with the IAEA, such as safety issues. But this co-operation, it should be emphasized, also makes sound political sense.

On 14 May, Gorbachev made a televised speech to the Soviet people concerning Chernobyl. He revealed that he was in favour of an "international mechanism" for the safe development of nuclear energy, which would include a system to warn other countries immediately in the event of an accident, particularly one regarding the release of radioactive materials. This apparent concession to those angered by Soviet silence about the accident was, however, accompanied by another political ploy: in the light of the Chernobyl disaster, said Gorbachev, "the Soviet Union has decided to extend its unilateral nuclear test moratorium to 6 August, the forty-first anniversary of Hiroshima."[106]

If Gorbachev had not made the connection between Chernobyl and Hiroshima clear enough, *Radio Moscow* elaborated further in a broadcast to North American listeners two days later, when Vladislav Koziakov was interviewed:

> *Interviewer*: Is there any connection between the Chernobyl accident and the Soviet decision to extend its moratorium?
> *Koziakov*: Most certainly. Moscow calls on Washington to look at the nuclear test ban more seriously now that the Chernobyl accident has shown once again what would happen to humanity in the case of a nuclear war.

What sort of "international mechanism" did Gorbachev have in mind? Boris Semenov felt that the USSR would support an international agreement for emergency help in the case of a nuclear accident, but not a binding agreement on safety standards for nuclear plants. Thus although the IAEA was to be encouraged to develop international guidelines, the actual implementation of these rules, in the Soviet view, should be left to the individual states.[107] The role of the IAEA was to be enhanced slightly, but it would have no binding role over any country, just as it had lacked such a role in the past. The stature of the IAEA grew considerably after the highly publicized visit of Hans Blix to the damaged reactor. But this did not signify that it was to have any genuine authority over the nuclear industry in the future.

After Chernobyl, the IAEA, which has 112 member states that rarely reach instant agreement on any issue, held meetings in May and June and put forward proposals for a modest increase in spending on safety at nuclear power plants, in addition to stressing the importance of international co-operation in the area of reliability and safety of atomic power plants (the latter represents virtually a perennial rather than a new proposal). The Soviet representative to the IAEA board, Boris Semenov, agreed to present a detailed account of the Chernobyl disaster to the IAEA in late August 1986.[108] The low-budget UN organization was therefore designated as the international medium for the future discussion of the accident.

Gorbachev and others subsequently repeated calls for an international system of safety measures. One such message was delivered verbally on Gorbachev's behalf to the UN Secretary General, Javier Perez de Cuellar, in New York City by the Soviet ambassador Iurii Dubinin in June. Gorbachev maintained that the IAEA and other international agencies should support attempts to develop new reactors with better safety features.[109] In July 1986, this line was developed further by Mikhail Ryzhov, another member of the Soviet delegation to the IAEA, who stated that the USSR supported the creation of a "safe nuclear reactor," which "should be developed an built under the auspices of the IAEA." According to Ryzhov, the Soviet authorities also wanted all IAEA members to provide information on the causes of nuclear breakdowns, on how they were combatted and on their ecological and radiological consequences.[110]

Was this an admission that Soviet reactors were unsafe? Later in July, Petrosiants revealed that "decisive and various" steps had been taken to increase safety at Soviet nuclear power plants, but the only specific improvement alluded to was more training for nuclear plant workers. Petrosiants did say, however, that some of the electrical equipment supplied to nuclear plants in the past was not of the highest quality, and that this quality would have to be checked more thoroughly in the future. Some nuclear plant workers "had forgotten what kind of fuel they were working with."[111]

The above statement was notably more frank than past Soviet comments. It should not, however, be taken too literally. Chernobyl may have enforced such belated *glasnost* [openness], but there is a major difference between the statement and the reality. Petrosiants was admitting what many in the industry knew already: that there were problems with equipment and with unskilled and inexperienced workers. Together with the disengagement of various safety mechanisms, the combination of the two had caused the Chernobyl disaster, as the Soviets had decided to reveal in Vienna in late August 1986. But Soviet plans had not changed. The question therefore is how the industry was to be revamped in the face of its unprecedented expansion, an expansion, moreover, that had been reconfirmed from various quarters after the accident? The IAEA was asked to construct a safe reactor, but before the organization could undertake such a task, what was to happen in the interim? No Soviet reactors, safe or unsafe, were to be dismantled. The IAEA's role is hence severely restricted by Soviet economic requirements that will not countenance any slowdown of the industry. The Soviet Union needs more electricity. Nuclear power is the source.

In conclusion, then, the Soviet nuclear industry has been plagued with problems from the first. It has been developed in spite of these difficulties as a means not only of supplying industries of the European part of the USSR with electricity, but also of co-ordinating a nuclear programme that also embraces the East European countries, using the most rapidly developing region in the sphere, Soviet Ukraine, as the medium. Most Soviet nuclear plants have experienced fundamental dilemmas that would have caused—it is fair to say—Western countries in similar positions to question the overall strategy. Soviet plants, let us say by Canadian standards, are unsafe. Chernobyl was unsafe. We have chronicled some of these problems above to show that they have intensified as the industry has expanded. The number of skilled workers has been spread too thinly, the supply and equipment questions have not been resolved, and, above all, the pace of development has been more rapid than warranted. Nuclear power has been perceived as a short-cut solution to a very real difficulty—securing what Gorbachev has termed the "scientific-technical acceleration" of the Soviet economy in the final years of the twentieth century.

Chernobyl, even in terms of long-term casualties, will not be the world's worst accident. It does not prove, as some have claimed, that the nuclear industry in the world as a whole, is inherently unsafe. It has demonstrated, however, that the Soviet nuclear industry has been operating in a manner that has bordered on the foolhardy. And at present, there have been few indications either that Chernobyl will change anything or that it will be the last such nuclear disaster.

Epilogue

The report delivered by the Soviet Government Commission to the International Atomic Energy Agency in Vienna was released in advance by the TASS news agency on 22 August 1986. In general, it confirms the general premise of this book that the Soviet approach to the development of nuclear energy has been fundamentally unsound.

An experiment had been authorized on the No. 8 turbo-generator at the Chernobyl nuclear power plant. But the conditions under which the experiment should have taken place were evidently violated by plant officials. With the fourth unit scheduled for routine maintenance work, the authorities at the plant wanted to find out how long electricity could be used from the turbo-generator while the reactor was being shut down. According to A. M. Petrosiants, the Chairman of the State Committee for the Utilization of Nuclear Energy, the experiment itself was not related to the reactor.

The plant officials, however, reportedly tried to take a short-cut in the experiment by preventing the automatic shutdown mechanisms of the reactor, which may have come into operation with the reactor operating at such a low capacity. Thus the reactor's emergency cooling system was shut off at 1400 hours on 25 April and remained shut down until the explosion occurred at 1.23 am the following morning. The system that safeguarded against an excessively low water level was also blocked off, and the operators evidently pulled out several of the reactor's control rods, also to stop an automatic shutdown. Extra pumps—more than the maximum permissible number—were turned on to raise the amount of steam flowing to the generator. The report stated that ''the operators were introducing disturbances almost constantly.''

By cutting off these vital safety mechanisms, a hazardous situation was created. The text of the report also confirmed (as was noted above) the basic instability of the RBMK 1000 reactor. The personnel involved in the 25 April experiment carried out their task without the approval of the reactor designer, the designer of the Chernobyl plant, or the personnel of the nuclear safety department at the plant.

In brief, the main reason for the disaster according to the Soviet account was a highly improbable combination of circumstances brought about by a barely credible negligence on the part of plant officials. Human error was therefore to blame. Yet, the experiment itself was neither unauthorized (it would have taken place at some point) nor the first to be conducted at the Chernobyl plant on the turbo-generators. In view of the punishments meted out by the authorities noted in Chapter Seven, and the nature of the experiment, it is clear that the Ministry of Power and Electrification of the USSR had a strong interest in the experiment's results. Whether or not this ministry ordered the experiment, it can be stated with certainty that the plant officials were not conducting experiments on their own initiative.

The full investigation of the Chernobyl disaster will likely take several years to complete. At this juncture, one can draw the following conclusions about the disaster, in the light of the Soviet Government Commission's report:

1. There is nothing in the report or elsewhere to suggest that an experiment on a commercial reactor was a unique or even unusual procedure. On the contrary, if proper safety mechanisms had been in place, the report implies, then such an accident would have been impossible.

2. The dismantling of at least six safety mechanisms at the fourth unit in order to conduct the experiment was the direct cause of the accident. The power surge from 7 to 50 per cent in a matter of seconds resulted from the loss of water coolant. In this respect, "human error" caused the Chernobyl disaster.

3. The use of graphite as a moderator on the RBMK 1000 reactor, and the lack of stability of that type of reactor generally were supplementary causes of the accident.

4. There was a major problem with labour discipline at the Chernobyl station. Gross violations of procedures occurred (regardless of whether those procedures were commendable in the first place).

5. The purpose of the experiment was to utilize electricity production to the maximum extent. This follows the current Soviet aim, outlined at the Twenty-Seventh Party Congress of the CC CPSU of March 1986, to conserve supplies of energy wherever possible. It was conducted therefore for economic rather than safety factors. This puts more onus for the disaster on the CPSU than on local plant officials, a point that the Soviet Government Commission has taken care to avoid.

The report also provided more information about the radiation levels around the plant after the explosion. Personnel at the plant reportedly received as much as 400 rems immediately, while on 27 April, within the 10-kilometre zone, an individual was receiving 1,000 millirems hourly. Within a short period, therefore, acute doses of radiation would have been incurred by those outdoors during that time. The level of radiation was much higher than Soviet reports had indicated previously, and the delay in evacuating Prypiat placed the health of citizens in jeopardy. One week after the accident, the radiation level in the city of Kiev, 150 kilometres to the south, was said to be eighty times above normal.

The prognosis for the return of evacuees in the report was considerably more pessimistic than earlier statements had indicated. A period of at least four years was anticipated before those moved from the plant vicinity would be allowed to return to their homes.

It should be acknowledged that the Soviet report, which ran over 300 pages in length, was relatively frank and open. Again, however, there have been few genuine indications that the accident will lead to any prolonged slowdown of the nuclear energy programme. The question that needs to be raised after Vienna remains the same: given the conditions under which Soviet nuclear power plants are built, the absence or very low levels of quality control, and the concomitant prime place assigned to nuclear power in the Soviet energy programme, can a disaster such as Chernobyl be attributed mainly (or solely) to human error? Was it not rather a consequence of both the way in which nuclear power plants are being organized and the current economic priorities of the USSR?

Notes

Notes to Chapter Two

1. *Robitnycha hazeta*, 11 November 1984.
2. I.F. Elliot, *The Soviet Energy Balance: Natural Gas, Other Fossil Fuels and Alternative Power Sources* (New York 1974), 123.
3. Cited by *UPI*, 19 January 1983.
4. *The Economist*, 19 September 1981.
5. *Izvestiia*, 4 March 1986.
6. *Visti z Ukrainy*, no. 30 (July 1985): 2.
7. See *Ekonomicheskaia gazeta*, no. 15 (April 1981): 2; and *Soviet Geography* (April 1981): 279.
8. *Izvestiia*, 4 December 1984.
9. *Radio Kiev*, 23 October 1985.
10. *Radio Moscow*, 15 December 1985.
11. *Pravda*, 30 June 1982.
12. *Soviet Geography* (December 1982): 770.
13. *TASS*, 25 September 1985.
14. *Sotsialisticheskaia industriia*, 9 March 1986.
15. *TASS*, 8 December 1981.
16. *Kazakhstanskaia pravda*, 7 October 1981.
17. *Pravda*, 4 October 1981.
18. *Soviet Geography* (December 1982): 769-71.
19. *Journal of Commerce*, 1 January 1983.
20. *TASS*, 3 October 1981.
21. *Pravda*, 30 June 1982.
22. *Radio Moscow*, 30 January 1986.
23. *Pravda*, 15 July 1982.
24. *Radio Moscow*, 11 February 1986.
25. *Meditsinskaia gazeta*, 5 February 1982.
26. *The Economist*, 27 March 1982.
27. *Narodnoe khoziaistvo SSSR v 1984 g.* (Moscow 1985), 166.
28. *Radio Moscow*, 6 September 1985.
29. *Moscow Television*, 6 September 1986.
30. *Radio Moscow*, 18 October 1985.
31. *Ibid.*, 29 October 1985.
32. *Sotsialisticheskaia industriia*, 12 February 1986.

NOTES TO CHAPTER 2

33. *Pravda*, 12 February 1986.
34. *Radio Moscow*, 1 March 1986.
35. *Radio Moscow* and *Novosti*, 4 March 1986.

Notes to Chapter Three

1. *RFE-RL* [*Radio Free Europe-Radio Liberty*] *Special* (Munich), 12 May 1986.
2. *Radio Prague*, 10 April 1985.
3. *Radio Sofia*, 22 May 1985.
4. *BTA* [Bulgarian news agency], 12 March 1986; and *RFE-RAD*, BR/72 (Ashley), 23 May 1986.
5. *BTA*, 12 March 1986.
6. *Rabotnichesko delo*, 10 February 1986.
7. *Tanjug* [Yugoslav news agency], 19 May 1986.
8. *Radio Prague*, 10 April 1985.
9. *CTK* [Czechoslovak news agency], 24 February 1985; and *Czechoslovak Television*, 3 February 1986.
10. *IAEA Report* (April 1986).
11. *Radio Prague*, 31 August 1985.
12. *Financial Times*, 6 September 1985.
13. *Radio Prague*, 17 October 1985.
14. *CTK*, 4 May 1986.
15. David Marples, "Comecon Cooperation in Nuclear Energy," *Soviet Analyst*, vol. 14, no. 15, 24 July 1985.
16. *MIT* [Hungarian news agency], 24 January 1986.
17. S. Koppany, "Hungary's First Nuclear Power Plant: A Monument to Inefficiency," *RFE-Hungarian Research*, 20 March 1986.
18. V. Socor, "Soviet-Romanian Programs in Nuclear Energy Development," *RFE-RAD Background Report/129*, 18 November 1985, 6.
19. *Agerpres* [Romanian news agency], 10 July 1985.
20. Reported by *Tanjug*, 11 February 1986.
21. *Tanjug*, 11 February 1986.
22. *RFE-RL Special*, 20 March 1985.
23. *Zycie Warszawy*, 13 January 1986.
24. *Ibid.*
25. *Ibid.*
26. *PAP* [Polish news agency], 15 November 1984.
27. *Journal of Commerce*, 1 May 1986.
28. *AP*, 12 April 1985.
29. Cited by *Reuter*, 1 May 1986.
30. *Journal of Commerce*, 1 May 1986.
31. *Tanjug*, 6 July 1985.
32. *Ibid.*, 27 February 1986.
33. *Ibid.*, 17 February 1986.
34. Cited in *Financial Times*, 5 March 1986.
35. *RFE-RAD* (de Weydenthal), 9 May 1986.
36. *Plan-Econ Report*, vol. 2, no. 19-20, 16 May 1986, p. 11.
37. *RFE-RAD* (Sobell and Matuska), 6 May 1986.
38. Cited in *RFE-RAD* (Sobell), 7 May 1986.
39. *CTK*, 14 May 1986.
40. *RFE-RAD* (Kusin), 20 May 1986.

41. *DPA* [West German news agency], 26 May 1986; and *AP*, 26 May 1986.
42. *Reuter*, 2 June 1986.
43. *Ibid.*, 3 June 1986.
44. *CTK*, 7 June 1986.
45. *Tygodnik Mazowsze*, no. 169, 8 May 1986, cited in *RFE-RAD PUE/8*, June 1986, 15.
46. *AP*, 9 May 1986.
47. *Ibid.*, 3 May 1986.
48. *PAP*, 15 May 1986.
49. *RFE-RL Central News Desk*, 17 May 1986.
50. *Kurier Polski*, 11 July 1986.
51. *AP, Reuter*, 16 May 1986.
52. *Ibid.*, 20 May 1986.
53. *Reuter, AP, DPA*, 1 June 1986.
54. *PAP*, 1 June 1986.
55. *Radio Warsaw*, 20 June 1986.
56. *BTA*, 29 May 1986.
57. See the account by Phillip Ashley, *RFE-RAD*, BR/72 (Ashley), 23 May 1986.
58. *Reuter*, 11 June 1986.
59. *Tanjug*, 6 May 1986.
60. *Ibid.*, 10 May 1986 [in English].
61. *Ibid.*, 10 May 1986 [in Serbo-Croatian].
62. Cited by *Reuter*, 12 May 1986.
63. *AP*, 26 May 1986.
64. Cited by *Reuter*, 11 July 1986.
65. *Radio Moscow*, 22 May 1986.
66. *Sotsialisticheskaia industriia*, 24 May 1986.
67. *Pravda*, 27 May 1986.

Notes to Chapter Four

1. *Energetika SSSR v 1981-1985 godakh* (Moscow 1981), 135-47.
2. *Ibid.*, 142.
3. *Ibid.*, 144.
4. *The New York Times*, 16 January 1983.
5. *Novosti*, 29 July 1983.
6. *Radio Moscow*, 14 May 1984.
7. *Moscow Television*, 26 December 1984.
8. *TASS*, 14 December 1984.
9. *Soviet Geography* (April 1985): 304.
10. *TASS*, 26 December 1985.
11. *Sotsialisticheskaia industriia*, 3 January 1986.
12. *Radio Moscow*, 3 March 1986.
13. *Sotsialisticheskaia industriia*, 9 March 1986.
14. *Izvestiia*, 5 March 1986.
15. *TASS*, 15 November 1985.
16. *Sotsialisticheskaia industriia*, 9 March 1986.
17. *Izvestiia*, 21 September 1984.
18. *Robitnycha hazeta*, 11 November 1984.
19. *Radianska Ukraina*, 24 March 1974.
20. *Soviet Geography* (May 1983): 393, 396.

21. D. Marples, "Recent Developments in the Ukraine's Nuclear Energy Program," *Radio Liberty Research Bulletin*, RL 405/85, 4 December 1985.
22. *Ekonomika Radianskoi Ukrainy*, no. 2 (February 1984): 14-16.
23. *Radianska Ukraina*, 30 July 1985.
24. *Radio Kiev*, 13 December 1984.
25. *Pid praporom leninizmu*, no. 2, January 1986.
26. *Radianska Ukraina*, 4 January 1986.
27. *TASS*, 31 January 1986.
28. *Pid praporom leninizmu*, no. 22, November 1984.
29. *Sovet Tojikistoni*, 15 November 1985.
30. Marples, "Recent Developments."
31. *Selskaia zhizn*, 9 September 1976.
32. *Radianska Ukraina*, 29 September 1973.
33. *Ibid.*, 29 December 1985.
34. *Radio Kiev*, 3 February 1986.
35. *Ibid.*
36. *Radianska Ukraina*, 4 January 1986.
37. *TASS*, 26 June 1986.
38. *Literaturna Ukraina*, 7 September 1976.
39. *Ibid.*, 28 September 1976. Eventually two hydro-electric power stations were built on this complex.
40. *Ibid.*, 3 August 1976.
41. *Ibid.*, 20 August 1976.
42. *Ibid.*, 19 August 1977.
43. *Ibid.*, 21 March 1978.
44. *Ibid.*, 19 August 1977.
45. *Ibid.*, 21 March 1978.
46. *Ibid.*, 19 August 1977.
47. *Ibid.*, 18 July 1978.
48. Cited in A. Kroncher, "Soviet Nuclear Power Station Construction Poses Constant Hazards," *Radio Liberty Research Bulletin*, RL 194/86, 15 May 1986.
49. Cited by George Stein in the *Los Angeles Times*, 16 May 1986.
50. *Literaturna Ukraina*, 18 July 1978.
51. *Ibid.*, 28 September 1976.
52. V. Socor, "Soviet-Romanian Progams in Nuclear Energy Development," *RFE-RAD*, Background Report/129, 18 November 1985.
53. *TASS*, 13 November 1985.
54. *Radio Moscow*, 18 March 1986.
55. *Robitnycha hazeta*, 26 July 1985.
56. *Izvestiia*, 20 June 1985.
57. *Sovetskaia Rossiia*, 16 September 1984.
58. *Radianska Ukraina*, 4 January 1986; and *Pid praporom leninizmu*, no. 2, January 1986.
59. *Izvestiia*, 5 April 1986.
60. *Radianska Ukraina*, 1 November 1985.
61. *Financial Times*, 10 May 1985.
62. *Robitnycha hazeta*, 11 October 1984.
63. *Radio Moscow*, 2 January 1986.
64. *Slowo Powszechne*, 13 March 1985.
65. *Ibid.*
66. *Robitnycha hazeta*, 9 April 1985.
67. *Ibid.*

68. *Slowo Powszechne*, 13 March 1985.
69. *Pid praporom leninizmu*, no. 2, January 1986.
70. *Radio Kiev*, 17 February 1986.
71. *Ibid.*, 19 March 1986.
72. *TASS*, 1 January 1985.
73. *Ibid.*, 30 December 1985.
74. *Radianska Ukraina*, 11 January 1986.
75. *Ibid.*
76. *Nashe slovo*, 16 February 1986.
77. *Radianska Ukraina*, 11 January 1986.
78. *Izvestiia*, 20 June 1985.
79. *Soviet Geography* (October 1985): 643, 646.

Notes to Chapter Five

1. *Energetika SSSR v 1981-1985 godakh* (Moscow 1981), 139.
2. *Novosti*, 29 July 1983.
3. *Radio Moscow*, 15 September 1983.
4. *Izvestiia*, 24 June 1984.
5. B. Semenov, "Nuclear Power in the Soviet Union," *IAEA Bulletin*, vol. 25, no. 2 (May 1986): 53-4.
6. *TASS*, 25 October 1984.
7. *Ibid.*, 21 February 1986.
8. *RFE-RL Special*, 29 August 1985.
9. S. Voronitsyn, "How Great a Problem is the Disposal of Nuclear Waste in the USSR?" *Radio Liberty Research Bulletin*, RL 500/81, 15 December 1981.
10. *Pravda*, 28 November 1981.
11. *Sotsialisticheskaia industriia*, 21 September 1983.
12. *Ibid.*
13. *Radio Moscow*, 1 April 1985.
14. *TASS*, 1 March 1986.
15. Semenov, *op. cit.*, 56-7.
16. *Reuter*, 29 August 1985.
17. *Literaturna Ukraina*, 14 June 1974.
18. *Sotsialisticheskaia industriia*, 21 September 1983.
19. *Literaturnaia gazeta*, 9 May 1984.
20. *Financial Times*, 10 May 1985.
21. *Kommunist*, no. 14 (September 1979): 25.
22. *Ibid.*, 26.
23. *Ibid.*
24. *Ibid.*, 27-8.
25. S. Voronitsyn, "Further Debate on the Safety of Nuclear Power Stations in the USSR," *Radio Liberty Research Bulletin*, RL 350/81, 7 September 1981.
26. *Visti z Ukrainy*, no. 2, January 1986.
27. *Frankfurter Rundschau*, 21 July 1983.
28. *Radio Moscow*, 22 January 1985.
29. *TASS*, 25 October 1984.
30. *Sotsialisticheskaia industriia*, 21 September 1983.
31. *TASS*, 31 January 1986.
32. *The Times*, 1 May 1986.
33. *The Economist*, 30 July 1983.

34. *TASS*, 19 July 1983.
35. *Pravda*, 20 July 1983.
36. *TASS*, 20 July 1983.
37. See for example, M. Haslett, "Soviet Nuclear Accident?" *BBC Current Affairs Talks*, 21 July 1983.
38. *Izvestiia*, 2 August 1983.
39. *Radio Moscow* [in Finnish to Finland], 15 August 1983.
40. *The Observer*, 14 August 1983.
41. *Pravda*, 15 September 1983.
42. *The New York Times*, 1 May 1986 (article by S. Diamond).
43. W. P. Geddes, "Nuclear Power in the Soviet Bloc," unpublished manuscript, Uranium Institute, London, May 1986.
44. *Sovetskaia Rossiia*, 30 May 1982.
45. *Moscow Television*, 10 February 1986.
46. *The New York Times*, 16 January 1983.
47. *Radio Moscow*, 30 July 1981.
48. *Sotsialisticheskaia industriia*, 8 December 1981.
49. For example, *Izvestiia*, 10 July 1979. Cited in A. Kroncher, "Soviet Nuclear Power Station Construction," *Radio Liberty Research Bulletin*, RL 194/86, 15 May 1986.
50. K. Bush, "Safety of Soviet Nuclear Power Plants Again Questioned," *Radio Liberty Research Bulletin*, RL 277/83, 21 July 1983.
51. *Komsomolskoe znamia*, 6 August 1985.
52. *Radio Vilnius* [in Russian], 1 November 1984.
53. *Daily Telegraph*, 21 May 1986.
54. A good survey of the problem is Kroncher, *op. cit.*
55. *Leningradskaia atomnaia elektrostantsiia imeni V.I. Lenina* (Moscow n.d.), 9; and *News From Ukraine*, no. 23, 1986.
56. *Daily Telegraph*, 30 April 1986.
57. *UPI*, 1 May 1986.
58. *Financial Times*, 30 April 1986.
59. *Wall Street Journal* (Europe), 15 May 1986.
60. *The Economist*, 3 May 1986.
61. *Radio Moscow*, 30 August 1984.
62. *Soviet Television/B*, 1605, 18 May 1986.
63. *Atomnaia energiia*, vol. 43, no. 6 (1977): 458.
64. Semenov, *op. cit.*, 51.
65. *Atomnaia energiia*, no. 6 (1977): 449.
66. J. DeBardeleben, "Esoteric Policy Debate: Nuclear Safety Issues in the USSR and GDR," unpublished manuscript, McGill University, Montreal, 1983, 9.
67. For example, *TASS*, 23 January 1985.
68. Geddes, 9.
69. *Radio Moscow*, 30 August 1984.
70. *Ibid.*, 22 March 1985.
71. *Ibid.*
72. Geddes, 12.
73. *AP*, 22 April 1979; *Financial Times*, 24 April 1979.
74. *RFE-RL Special*, 21 July 1983.
75. *Soviet Analyst*, vol. 15, no. 9 (30 April 1986): 2.
76. *Der Spiegel*, 25 July 1983; *Baltimore Sun*, 2 April 1979; *Soviet Analyst* (30 April 1986): 2.
77. *Der Spiegel*, 25 July 1983.
78. *The Globe and Mail* (Toronto), 22 May 1986.

Notes to Chapter Six

1. *Istoriia mist i sil Ukrainskoi RSR: Kyivska oblast* (Kiev 1971), 9.
2. *Nauka i suspilstvo* (November 1971).
3. *Ibid.*
4. *Literaturna Ukraina*, 23 July 1976; and *Radianska Ukraina*, 29 October 1977.
5. *Boston Herald*, 10 May 1986.
6. *RFE-RL Special*, 29 April 1986; and *Financial Times*, 30 April 1986.
7. *Radio Kiev*, 20 February 1986.
8. *Reuter*, 14 May 1986.
9. *Soviet Life* (February 1986): 13.
10. *Literaturna Ukraina*, 14 June 1974.
11. *Ibid.*
12. *Ibid.*
13. *Ibid.*, 23 July 1976.
14. *Radianska Ukraina*, 29 October 1977.
15. *Ibid.*
16. *AP*, 1 May 1986.
17. *Pravda*, 26 July 1985.
18. *Radianska Ukraina*, 29 December 1985.
19. *Vitchyzna* (March 1986): 155-61.
20. *Literaturna Ukraina*, 27 March 1986, and ff.
21. *Interview C*, AECL, 29 July 1986.
22. At the time of writing, the Soviet authorities had agreed to give a "full account" of the causes of the accident to a meeting of the International Atomic Energy Agency in Vienna in late August.
23. *Daily Telegraph*, 30 April 1986.
24. Cited by *Reuter*, 9 June 1986.
25. *Radio Kiev*, 7 May 1986.
26. *Radio Budapest*, 10 May 1986.
27. *Novosti* [in English], 12 May 1986.
28. Cited in the *Washington Post*, 22 May 1986.
29. *RFE-RL Special*, 22 May 1986.
30. For example, *The Observer*, 25 May 1986; and *CBC The Journal*, 24 June 1986.
31. *RFE-RL Special*, 27 May 1986.
32. *Interview D*, AECL, 30 July 1986.
33. *Interview A*, AECL, 29 July 1986; *News From Ukraine*, no. 23, 1986.
34. According to interviewee B, of AECL, 8 August 1986, this itself would explain very little. The turbine would soon stop spinning if the reactor was no longer producing heat. In the RBMK reactor (in contrast to the CANDU, for example), the same water passes through the reactor to the turbine, and it is conceivable that the flow might have been interrupted in an attempt to keep the turbine in motion. Would the purpose have been to conserve power in the production of electricity?
35. *Interviews A and C*, AECL, 29 and 30 July 1986.
36. *Pravda Ukrainy*, 8 May 1986.
37. *Ekonomicheskaia gazeta*, no. 23 (June 1986).
38. *Pravda Ukrainy*, 8 May 1986.
39. *Soviet News and Views* [Ottawa: USSR Embassy Press Office] (August 1986), 4.
40. *Izvestiia*, 10 May 1986.
41. *Ibid.*, 19 May 1986.
42. *Ibid.*
43. *Ekonomicheskaia gazeta*, no. 23 (June 1986).

44. *Izvestiia*, 10 May 1986.
45. *Komsomolskaia pravda*, 16 May 1986.
46. *Nashe slovo*, 29 June 1986; and *Ekonomicheskaia gazeta*, no. 23 (June 1986).
47. *Interview G*, AECL, 29 July 1986.
48. *Radiation Is Part Of Your Life*, Atomic Energy of Canada Limited (Pinawa, Manitoba 1983), 16.
49. *Ibid.*, 22.
50. E.J. Hall, *Radiation and Life* (Oxford 1980), 48.
51. *Radiation Is Part*, 18.
52. Semenov, *op. cit.*, 54.
53. *UPI*, 15 May 1986.
54. *Krasnaia zvezda*, 13 May 1986.
55. *AP*, 16 May 1986.
56. *Soviet Television*, 1605 hours, 18 May 1986.
57. *Reuter*, 9 May 1986.
58. *Nashe slovo*, 29 June 1986.
59. *Pravda*, 4 June 1986.
60. *Soviet Television*, 1940 hours, 5 June 1986.
61. *AP*, 3 May 1986.
62. *Financial Times*, 3 July 1986.
63. *RFE-RL Special*, 29 June 1986.
64. *TASS*, 10 June 1986.
65. *Literaturna gazeta*, 21 May 1986.
66. *Izvestiia*, 28 May 1986.
67. *Ibid.*
68. *Radio Moscow*, 6 May 1986; *Novosti*, 12 May 1986; and *TASS*, 13 May 1986.
69. *Novosti*, 12 May 1986.
70. *Izvestiia*, 28 May 1986.
71. Richard Champlin in the *Washington Post*, 13 July 1986.
72. *AP*, 6 June 1986.
73. *Sunday Times*, 29 June 1986.
74. *Washington Post*, 13 July 1986.
75. *Soviet Television*, 1630 hours, 3 June 1986.
76. Robert P. Gale on *CBC The Journal*, 24 June 1986.
77. *Sovetskaia Rossiia*, 29 June 1986.
78. *Washington Post*, 2 July 1986.
79. *Komsomolskaia pravda*, 16 May 1986.
80. *Reuter*, 26 May 1986.
81. *CBC News*, 2 August 1986.
82. *AP*, 14 June 1986.
83. *Pravda*, 20 May 1986.
84. *Reuter*, 9 May 1986.
85. *Radio Budapest*, 10 May 1986.
86. *Ekonomicheskaia gazeta*, no. 23 (June 1986).
87. V. Tolz, "USSR Grudgingly Discloses More Facts About Chernobyl," *Radio Liberty Research Bulletin*, RL 187/86, 14 May 1986.
88. *CBC The Journal*, 24 June 1986.
89. *Pravda*, 3 June 1986.
90. *Pravda Ukrainy*, 13 May 1986.
91. *AFP*, 9 May 1986.
92. *AP*, 5 July 1986.
93. *Pravda*, 12 May 1986.

94. *Izvestiia*, 20 May 1986.
95. *Trud*, 11 June 1986.
96. *Pravda Ukrainy*, 13 May 1986.
97. *Izvestiia*, 20 May 1986.
98. *Sovetskaia Rossiia*, 10 June 1986.
99. *Izvestiia*, 14 June 1986.
100. *Radio Moscow*, 21 May 1986.
101. *Pravda*, 21 May 1986.
102. *Ibid.*, 5 June 1986.
103. *Izvestiia*, 20 May 1986.
104. *Pravda*, 5 July 1986.
105. *RFE-RAD* (Girnius), 1 July 1986.
106. *AP*, 5 July 1986.
107. *Pravda*, 21 May 1986.
108. *Radio Moscow*, 7 May 1986.
109. *Radio Kiev*, 5 and 6 May 1986.
110. *Ibid.*, 9 May 1986.
111. *Pravda Ukrainy*, 11 May 1986.
112. *Pravda*, 18 May 1986.
113. *Sovetskaia Rossiia*, 20 May 1986.
114. *Pravda*, 18 May 1986.
115. *Reuter*, 9 May 1986.
116. *Pravda*, 9 May 1986.
117. *The Times*, 7 June 1986.
118. See, for example, *Radio Moscow*'s broadcast of 10 May 1986 about a May-Day visitor from a British trade-union organization who was said to be "extremely indignant about the grossly exaggerated reaction of governments and certain organizations in the West over the accident at Chernobyl."
119. *UPI*, 8 June 1986.
120. *Pravda Ukrainy*, 11 May 1986.
121. *Izvestiia*, 28 May 1986.
122. *Sovetskaia Rossiia*, 10 June 1986.
123. *Komsomolskaia pravda*, 18 July 1986.
124. *PAP*, 27 May 1986.
125. *Pravda Ukrainy*, 1 July 1986.
126. *Radio Moscow*, 24 May 1986.
127. *Literaturnaia gazeta*, 21 May 1986.
128. The following are based on conversations between Soviet and emigre Ukrainians that have taken place since the Chernobyl disaster.

Notes to Chapter Seven

1. This is untrue. The accident did occur two weeks earlier, but it was not reported in newspapers until 29 April in Ukraine, and 30 April elsewhere in the USSR. Thus at the most, the first coverage preceded the quoted article by only 10 days.
2. S. Kondrashov, "Thinking About Chernobyl," *Izvestiia*, 9 May 1986.
3. *Izvestiia*, 10 August 1986.
4. *Pravda Ukrainy*, 8 May 1986.
5. *Radio Moscow*, 24 May 1986.
6. *Sovetskaia Rossiia*, 10 June 1986.
7. *Pravda*, 20 May 1986.

8. *Sovetskaia Rossiia*, 10 June 1986.
9. *Izvestiia*, 20 May 1986.
10. *Krasnaia zvezda*, 7 June 1986; and 16 May 1986.
11. *Soviet Television*, 1630 hours, 12 May 1986.
12. *Ekonomicheskaia gazeta*, no. 23 (June 1986).
13. *Soviet Television*, 1630 hours, 12 May 1986.
14. *Krasnaia zvezda*, 7 June 1986.
15. *Sovetskaia Rossiia*, 18 May 1986; *Pravda*, 13 May 1986.
16. *Radio Moscow*, 12 May 1986.
17. *TASS*, 15 May 1986.
18. *Robitnycha hazeta*, 28 June 1986.
19. *Krasnaia zvezda*, 18 May 1986.
20. *Robitnycha hazeta*, 5 June 1986.
21. *Tanjug*, 13 May 1986; and *TASS*, 6 June 1986.
22. *TASS*, 6 June 1986.
23. *Robitnycha hazeta*, 5 June 1986.
24. *Soviet Television*, 1900 hours, 7 June 1986.
25. *Komsomolskaia pravda*, 4 July 1986.
26. *Pravda*, 13 May 1986.
27. *Radio Moscow*, 30 May 1986.
28. *TASS*, 15 May 1986.
29. *Sovetskaia Rossiia*, 14 May 1986.
30. *Trud*, 25 May 1986.
31. *Izvestiia*, 20 May 1986.
32. *Reuter*, 25 May 1986.
33. *TASS*, 20 July 1986.
34. *Radio Moscow*, 27 May 1986.
35. *Komsomolskaia pravda*, 6 June 1986.
36. *Pravda*, 4 June 1986.
37. Cited by Martin Walker of *The Guardian*, 18 June 1986. See also *Pravda*, 4 June 1986; and *UPI*, 14 June 1986.
38. *Selskaia gazeta*, 25 May 1986.
39. See "Reservists From Latvia Help At Chernobyl," *RFE-RAD* (Bungs), 10 July 1986.
40. *Rabochaia gazeta*, 28 June 1986.
41. *TASS*, 8 July 1986.
42. *Ibid.*
43. *Pravda*, 19 May 1986.
44. *Radio Moscow*, [1200 hours] 4 June 1986.
45. *Ibid.*, [1100 hours] 4 June 1986.
46. *Soviet Television*, 1630 hours, 5 June 1986.
47. *RFE-RL Central News Desk*, 13 May 1986.
48. *Soviet Television*, 1900 hours, 17 May 1986.
49. *Pravda*, 27 June 1986.
50. *Radio Moscow*, 18 May 1986.
51. *Izvestiia*, 17 May 1986.
52. *Soviet Television*, 1900 hours, 7 June 1986.
53. *Radio Moscow*, 20 May 1986.
54. *Ibid.*, 19 May 1986.
55. *Krasnaia zvezda*, 25 May 1986.
56. *Radio Moscow*, 28 May 1986.
57. *Izvestiia*, 17 May 1986.

58. *Ibid.*
59. Bungs, *op. cit.*, 10 July 1986.
60. *Rabochaia gazeta*, 11 June 1986.
61. *Krasnaia zvezda*, 1 June 1986.
62. *Radio Moscow*, 24 May 1986.
63. *Der Spiegel*, 11 May 1986.
64. R. Solchanyk, "Chernobyl': The Political Fallout in Kiev," *Radio Liberty Research Bulletin*, RL 182/86, 5 May 1986.
65. *Pravda*, 18 May 1986.
66. *AP*, 19 May 1986.
67. *Novosti*, 26 May 1986.
68. Solchanyk, *op. cit.*
69. See, for example, the *Christian Science Monitor* article of 14 May 1986 by Donald Van Atta. A good analysis of the problems involved in such speculations is provided in R. Solchanyk, "The Perils of Prognostication," *Soviet Analyst*, vol. 15, no. 5, 5 March 1986.
70. *Radio Kiev*, 20 May 1986.
71. *Pravda Ukrainy*, 3 July 1986.
72. *Radio Moscow*, 4 July 1986.
73. *Ibid.*, 11 July 1986.
74. See D.R. Marples, "Crisis in Soviet Industry? An Examination of the Soviet Steel Industry in the 1980s," forthcoming, *Canadian Slavonic Papers* (December 1986).
75. D.R. Marples, "Retirement of First Secretary of Kiev Oblast Party Committee," *Radio Liberty* (F-590), 6 November 1985. See also Solchanyk, *op. cit.*, 5 May 1986.
76. *Komsomolskaia pravda*, 17 May 1986.
77. Cited in *AP*, 17 May 1986.
78. Cited in *ibid.*, 31 May 1986.
79. *Pravda*, 3 June 1986.
80. *Ibid.*, 15 June 1986.
81. *Novosti* [in English], 26 May 1986.
82. *RFE-RL Special*, 27 May 1986.
83. *Pravda*, 3 June 1986.
84. *TASS*, 20 July 1986.
85. *Pravda*, 15 June 1986.
86. *Pravda Ukrainy*, 11 June 1986.
87. *TASS*, 20 July 1986; and *Edmonton Journal*, 31 July 1986.
88. *Reuter*, 25 May 1986.
89. *Pravda*, 3 June 1986.
90. *Soviet Television*, 5 June 1986.
91. *Radio Moscow*, 10 July 1986.
92. *Pravda*, 9 July 1986.
93. Cited in *Edmonton Journal*, 19 July 1986.
94. *Izvestiia*, 4 August 1986.
95. *Radio Kiev*, 14 July 1986.
96. *Izvestiia*, 15 July 1986.
97. *Pravda*, 3 June 1986.
98. *News From Ukraine*, no. 30, 1986; *Robitnycha hazeta*, 8 August 1986; *Izvestiia*, 10 August 1986.
99. *Pravda*, 26 May 1986; *Soviet Television*, 2 June 1986.
100. *Radio Budapest*, 15 May 1986.
101. *Ekonomicheskaia gazeta*, no. 23 (June 1986).
102. *Novosti*, 26 May 1986.

103. *Pravda*, 2 June 1986.
104. *Soviet Television*, 5 June 1986.
105. *Pravda Ukrainy*, 19 June 1986.
106. *Soviet Television*, 14 May 1986.
107. *RFE-RL Special*, 21 May 1986.
108. This book went to press before the IAEA meeting in Vienna on 25 August.
109. *AP*, 3 June 1986.
110. *TASS*, 7 July 1986.
111. *Pravda*, 31 July 1986.

Selected References

AFP (French press agency)

Agerpres (Romanian press agency)

Associated Press (AP)

Atomic Energy Canada Ltd. *Radiation Is Part Of Your Life* (Pinawa, Manitoba 1983).

Atomnaia energiia

BBC Current Affairs Talks

BTA (Bulgarian news agency, Sofia)

Baltimore Sun

Borsa, Joseph. "Radiation and Public Health—Fact Versus Myth," (Atomic Energy of Canada Limited, Pinawa, Manitoba 1983).

Boston Herald

Bungs, Dzintra. "Reservists From Latvia Help at Chernobyl'," *RFE-RAD*, 10 July 1986.

Bush, Keith. "Safety of Nuclear Power Plants Again Questioned," *Radio Liberty Research Bulletin*, RL 277/83, 21 July 1983.

CBC News

CBC The Journal

CTK (Czechoslovak news agency, Prague)

Canadian Slavonic Papers

"Chernobyl's Repercussions For Poland," *RFE-Polish SR/10* (L.V.), 27 June 1986.

Czechoslovak Television

The Economist

DPA (West German news agency)

Daily Telegraph

DeBardeleben, J., "Esoteric Policy Debate: Nuclear Safety Issues in the USSR and GDR," unpublished manuscript, McGill University, Montreal, 1983.
Der Spiegel
Edmonton Journal
Ekonomicheskaia gazeta
Ekonomika Radianskoi Ukrainy
Elliot, I. F., *The Soviet Energy Balance: Natural Gas, Other Fossil Fuels and Alternative Power Sources* (New York 1974).
Energetika SSSR v 1981-1985 godakh (Moscow 1981).
Fetter, Steven A. and Kosta Tsipis. "Catastrophic Releases of Radioactivity," *Scientific American* (April 1981): 41-7.
Financial Times
Frankfurter Rundschau
Geddes, W. P., "Nuclear Power in the Soviet Bloc," unpublished manuscript, Uranium Institute, London, May 1986.
The Globe and Mail
The Guardian
Hall, E.J. *Radiation And Life* (Oxford 1980).
Haslett, Malcolm. "Soviet Nuclear Accident?" *BBC Current Affairs Talks*, 21 July 1983.
IAEA [International Atomic Energy Agency] Bulletin
Istoriia mist i sil Ukrainskoi RSR: Kyivska Oblast (Kiev 1971).
Izvestiia
Journal of Commerce
Kazakhstanskaia pravda
Kommunist
Komsomolskoe znamya
Koppany, S. "Hungary's First Nuclear Power Plant: A Monument To Inefficiency," *RFE-Hungarian Research*, 20 March 1986.
Krasnaia zvezda
Kroncher, A. "Soviet Nuclear Power Station Construction Poses Constant Hazards," *Radio Liberty Research Bulletin*, RL 194/86, 15 May 1986.
Krutikov, P.G. and V.V. Chepkunov. *Leningradskaia AES* (Leningrad 1984).
Kurier Polski
Leningradskaia atomnaia elektrostantsiia imeni V.I. Lenina (Moscow, n.d.).
Literaturna Ukraina
Literaturnaia gazeta
Los Angeles Times
Marples, David. "Comecon Cooperation in Nuclear Energy," *Soviet*

Analyst, vol. 14, no. 15, 24 July 1985.

Marples, David. "Nuclear Energy in the Ukraine: The Wave of the Future," *Radio Liberty Research Bulletin*, RL 451/84, 27 November 1984.

Marples, David. "Recent Developments in the Ukraine's Nuclear Energy Program," *Radio Liberty Research Bulletin*, RL 405/85, 4 December 1985.

MIT (Hungarian news agency, Budapest)

Meditsinskaia gazeta

Moscow Television

Narodnoe khoziaistvo SSR v 1984g (Moscow 1985).

Nashe slovo

Nauka i suspilstvo

The New York Times

News From Ukraine

Novosti (Soviet news agency)

Nuclear Engineering International

"Nuclear Power, the Environment and Man," (IAEA, Vienna, 1984).

The Observer

PAP (Polish news agency, Warsaw)

Pid praporom leninizmu

Plan-Econ Report (Washington, D.C.)

Pravda

Pravda Ukrainy

Die Presse (Vienna)

The Programme of the Communist Party of the Soviet Union (Moscow 1986).

Rabotnichesko delo

Radianska Ukraina

Radio Budapest

Radio Free Europe Research

Radio Kiev

Radio Liberty Research Bulletin

Radio Moscow

Radio Prague

Radio Sofia

Radio Vilnius

Radio Warsaw

Reuter

Robitnycha hazeta [Russian: *Rabochaia gazeta*]

Rogers, J.T. "Candu Moderator Provides Ultimate Heat Sink in a LOCA," *Nuclear Engineering International* (January 1979): 38-41.

Ryzhkov, Nikolai. *Guidelines for the Economic and Social Development of the USSR for 1986-1990 and for the Period Ending in 2000* (Moscow 1986).

Scientific American

Selskaia gazeta

Selskaia zhizn

Semenov, B.A. "Nuclear Power in the Soviet Union," *IAEA Bulletin*, vol. 25, no. 2 (May 1986): 47-59.

Silski visti

Slowo Powszechne

Socor, Vladimir. "Soviet-Romanian Programs in Nuclear Energy Development," *RFE-RAD Background Report*/129, 18 November 1985.

Solchanyk, Roman. "Chernobyl': The Political Fallout in Kiev," *Radio Liberty Research Bulletin*, RL 182/86, 5 May 1986.

Solchanyk, Roman. "The Perils of Prognostication," *Soviet Analyst*, vol. 15, no. 5, 5 March 1986.

Solidarnosc News (Toronto)

Sotsialisticheskaia industriia

Sovet Tojikistoni

Sovetskaia Rossiia

Soviet Analyst

Soviet Geography

Soviet Life

Soviet Nationality Survey

Soviet News and Views

Sunday Times

TASS

Tanjug (Yugoslav news agency, Belgrade)

The Times

Tolz, Vera. "The Chernobyl Commission," *Radio Liberty Research Bulletin*, RL 275/86, 15 July 1986.

Tolz, Vera. "USSR Grudgingly Discloses More Facts About Chernobyl," *Radio Liberty Research Bulletin*, RL 187/86, 14 May 1986.

Trud

United Press International (UPI)

Visti z Ukrainy

Vitchyzna

Voronitsyn, S. "How Great a Problem is the Disposal of Nuclear Waste in the USSR?" *Radio Liberty Research Bulletin*, RL 500/81, 15 December 1981.

Voronitsyn, S. "Further Debate on the Safety of Nuclear Power Stations

in the USSR," *Radio Liberty Research Bulletin*, RL 350/81, 7
September 1981.
Wall Street Journal
Washington Post
Znannia iunosti
Znannia ta pratsia (Kiev)
Zycie Warszawy

Appendix 1

OFFICIALS REPRIMANDED AFTER CHERNOBYL ACCIDENT

12 May:

Dismissals from their posts of three party members who worked at Chernobyl nuclear power plant: O. Shapoval, an engineer with the Chernobyl subsidiary of the South Energy Construction division, who was expelled from the party; A. Sichkarenko, another leader of this same collective, who was severely reprimanded and had his party card endorsed by the Prypiat urban party committee; and O.Hubskii. A deputy head of the youth section of the construction department, Iurii Zahalsky, was also removed from his position for "shirking his duties."

3 June:

Reported that 177 party members were still unaccounted for after the disaster. Attack on Ministry of Power and Electrification of the USSR by party leaders of Prypiat and Chernobyl mounts.

15 June:

A meeting of Kiev oblast party committee dismisses Chernobyl nuclear power plant Director, V. Briukhanov, and Chief Engineer, N. Fomin for alleged failure to make a correct assessment of the accident and to take the necessary measures. Also reprimanded Deputy Director, R. Soloviev, for abandoning his post, and two other Deputy Directors, I. Tsarenko and V. Hundar, for failing their duties. E. Pozdyshev named as the new Director of the Chernobyl plant.

20 July:

CC CPSU Politburo issues statement announcing the dismissals of First Deputy Minister of Power and Electrification of the USSR, G. Shasharin and severely warns Minister of Power and Electrification of the USSR, A. Maiorets. Dismissal of Evgenii Kulov, Chairman of the State Committee for the Supervision of the Safe Working Practices in the Atomic Energy Industry. Both ministries said to be guilty of gross negligence and lack of control over the nuclear power plant. Dismissal of Ivan Emelianov, well known nuclear expert who was involved in the design of the Chernobyl plant. Emelianov was a Corresponding Member of the Academy of Sciences of the USSR and Deputy Director of the Institute for Energy Equipment Studies. The statement also declared that the former Director of the Chernobyl plant, V. Briukhanov, had been expelled from the Communist Party.

31 July:

Former Chief Engineer at Chernobyl, N. Fomin, is expelled from the party.

14 August:

Six party members disciplined. Aleksei Makukhin, First Deputy Minister of Power and Electrification of the USSR is severely reprimanded for failing to take measures to improve the reliability of the Chernobyl plant. Viktor Sydorenko, Deputy Chairman of the State Committee for the Supervision of Safe Working Practices in the Atomic Energy Industry, and Deputy Chairman, M.P. Alekseiev, are severely reprimanded for their failure to prevent violations of the safety regulations at the Chernobyl plant. L.P. Mikhailov, Director of the Hydroproject Institute of the Ministry of Power and Electrification of the USSR is reprimanded for not ensuring the safe conducting of the generator test at Chernobyl. G. Veretennikov, head of the Union of Atomic Energy industrial association (Ministry of Power and Electrification) and Ie. V. Kulikov, of the Ministry of Medium Machine-Building of the USSR are expelled from the Communist Party for their alleged weak leadership and lack of discipline in their work.

Appendix 2

"CHERNOBYL IS A WARNING"

The following statement by the TASS news agency about the causes of the accident appeared in the newspaper Izvestiia *on 23 August 1986. The translation is the author's.*

On 21 August, Soviet and foreign journalists received concise information about the reasons for the accident at the Chernobyl atomic energy station and its consequences. Speaking at a meeting in the press centre of the MID USSR, A.M. Petrosiants, the Chairman of the State Committee for the Utilization of Atomic Energy, reported that detailed information about the accident and its consequences has been presented to the International Atomic Energy Agency (IAEA) and will be discussed at a meeting of technical experts from member-countries of the IAEA on 25 August.

What was the cause of the dramatic event that occurred at Chernobyl on 26 April? A.M. Petrosiants responded to this question:

In the decree of the Central Committee of the Communist Party of the Soviet Union, based on the report of the Government Commission, it was noted that the accident occurred because of a whole series of violations of the operative rules by workers at the atomic energy station. The fourth generating unit of the station came on-line in December 1983, and worked, like all the other units, in a totally satisfactory fashion. In 1985, for example, the station produced 29 billion kilowatt/hours of electricity with an efficient use of its capacity. Perhaps the uninterrupted performance of the station led to a certain complacency, a matter-of-fact attitude. And this could have been an indirect cause of the irresponsibility, negligence and lack of discipline which, in the final analysis, led to grave consequences.

Before the shutting down of the fourth generating unit for regular maintenance after two years of operation, a test was carried out on one of the turbo-generators. The goal of the test formed part of an experiment that was investigating the opportunities for using the mechanical energy of the turbo-generator rotor to maintain the operation of mechanisms of this unit under secure conditions.

During the investigation, it was established that the quality of the test programme was low, and that necessary safety precautions were not anticipated. On 25 April, at 1400 hours, the emergency cooling system of the reactor was shut off, and those performing the experiment already wanted to begin the tests, but on the orders of the Controller of the Kiev Energy Association, the shutting down of the reactor was delayed, and it continued to operate until 1.23 am, i.e., to the moment of the accident, with its emergency cooling system shut down.

The programme for the test had not been approved either with the representatives of the reactor constructors, or with the chief planners of the station, or with the scientists of the nuclear safety committee who are constantly stationed at the plant. The culprits behind the accident have been severely reprimanded, emphasized A.M. Petrosiants, but for a long time, the lesson of Chernobyl will remind us of the necessity of strict, careful, attention to technology in general, and to new technology in particular.

The Chernobyl event focused attention on today's question concerning the guaranteeing of the international safety of nuclear energy. On the initiative of the USSR, the IAEA, with the participation of the USSR and other member-states, is organizing two conventions: "Concerning Compulsory Notification about Nuclear Accidents where Occurring" and "Concerning Aid to Countries in the Case of an Accident at an Atomic Energy Station." Both these conventions will be examined at special international conferences of the IAEA in Vienna at the end of September.

V.A. Legasov, the First Deputy Director of the Kurchatov Atomic Energy Institute of the Academy of Sciences of the USSR, informed journalists that "We consider that by sharing with other countries that to which we were exposed, all our colleagues who use nuclear energy will find it useful and be able to adopt a critical and constructive conclusion. We are prepared also to receive information from technical experts of the IAEA member-states at the forthcoming conference directed toward improving the effectiveness of deactivization work.

Academician V.A. Legasov reported that the presentation to the IAEA of the causes of the accident consists of two folios. The first summarizes the report and includes numerous data, measurements and observations that were taken at the moment of the accident and directly after it. The second is 350-pages long. It contains factual material about the construction of the reactor, its technical characteristics, and the ecological and medical aspects of the problem.

L.A. Ilin, the Vice-President of the Academy of Medical Sciences of the USSR, declared that the state of health of those people given medical attention after the accident is improving, and that many of them are recuperating in sanatoriums and rest homes. However, all these people will be re-examined medically at regular intervals. To co-ordinate and to direct this work, to carry out further research, Centre for Radiological Medicine with the Academy of Medical Sciences of the USSR has been created in Kiev.

Speaking about the consequences of the accident for the natural environment, Iu.A. Izrael, the head of the State Committee for Hydrometeorology of the USSR, noted that in the zone adjacent to the atomic energy station, in certain polluted places, there are radioactive particles on the surface, but their penetration into the soil has not exceeded a few millimetres. Concerning the situation of the Dnieper water basin, which has special importance, the level of radioactivity, even in the Prypiat River, was very insignificant. Concerning this question, certain special measures were applied. At present, the composition of the water in the Dnieper basin is being controlled constantly. Strict dosimetric control is carried out of all the products coming into the stores and onto the markets.

The journalists were interested in how the situation at the fourth generating unit is being controlled today.

"The situation is controlled with a whole series of special measures created specifically for this system," responded Academician V.A. Legasov.

"When will the other reactors be brought into operation?" "In the Fall of this year, the first and second generating units of the atomic energy station will be allowed to resume operations."

The correspondent of the Bulgarian newspaper *Rabotnichesko delo* asked how the Soviet initiatives in the area of the safe development of nuclear energy had been received. The head of the section of the MID USSR for the Peaceful Utilization of Nuclear Energy and the Cosmos, Iu.K. Nazarkin, stressed that the proposals put forward by M.S. Gorbachev in May of this year had received world-wide support. The plans of the convention, which were elaborated at a recent meeting in Vienna, could become the basic principles of a system for the safe development of nuclear energy. The Soviet Union, like the overwhelming majority of the participants at the Vienna meeting, proposes that proper notification be given in the event of any nuclear accident, whether at a peaceful installa-

207

tion or at a military site, including accidents related to nuclear arms testing. This suggestion was declared unacceptable only by the United States. However, it was agreed [by all parties] that notification concerning accidents in nuclear arms and related to nuclear tests would be carried out on a voluntary basis. The preparation for the conventions on notification and aid in cases of an accident are important steps in the creation of an international regime for the safe development of nuclear energy.

Thus the accident, the details of which journalists have received today, is a warning. And not only for specialists and experts in nuclear power and energy questions. This event, as was underlined in the television address of General Secretary of the CC CPSU, M.S. Gorbachev, has provided an object-lesson of what would occur on a much worse scale with nuclear arms conflagration. One barrier against this menace is the unilateral moratorium of the Soviet Union on nuclear arms testing, which was extended once again on 18 August.

Index of Personnel

Adelmann, Kenneth.
U.S.arms negotiator.
137.
Akimov, A.
Shift supervisor, fourth generating unit, Chernobyl nuclear power plant.
140.
Ananenko, A.
Twenty-seven-year old engineer at the second generating unit of Chernobyl nuclear power plant. Volunteered to empty the bubbler pond under the reactor after the accident.
157.
Anderson, Jack.
U.S. journalist.
96.
Andropov, Iurii.
General Secretary of the CC CPSU and former KGB chief. Died in 1984.
90.
Antifeev, T.I.
Supervisor in the Construction Department, South Ukraine nuclear power plant.
81.
Antonov, Aleksei.
Chairman of the CMEA Intergovernment Commission for Atomic Energy Equipment.
53, 55.
Antoshchkin, Nikolai.
Air-force Major-General. Commanded the helicopter pilots entrusted with plugging the damaged reactor from the air after the accident.
156.

Arbatov, Georgii.
Head of the Soviet Institute for the Study of United States and Canada.
34.
Baranov, A.Ia.
Radiation specialist at Moscow's Hospital No. 6. Performed bone marrow transplants on severely affected victims of Chernobyl accident.
138, 140.
Baranov, B.
Shift supervisor, Chernobyl nuclear power plant. Volunteered to empty bubbler pond along with Ananenko.
157.
Belitsky, Boris.
Host of a science and engineering programme on *Radio Moscow*.
2, 96, 98.
Bener, A.I.
Fire chief at Odessa nuclear power and heating plant.
86.
Berdov, G.V.
Deputy Chief of the Ukrainian MVD.
32, 128, 155.
Bespalov, V.
Engineer at Chernobyl nuclear power plant. Volunteered to empty bubbler pond along with Ananenko and Baranov.
157.
Bibin, L.
First Deputy Chairman, USSR State Planning Committee.
73-4.
Blix, Hans.
Director-General, International Atomic Energy Agency.
30, 141, 178.
Bratchenko, Borys. Minister of the Coal Industry of the USSR until dismissal in December 1985.
40.
Brezhnev, Leonid.
General Secretary of the CC CPSU and Soviet President. Died in 1982.
24, 73, 105, 168.
Brezhnev, Vladimir.
Minister of Transport Construction of the USSR.
165.
Briukhanov, V.
Director of Chernobyl nuclear power plant. Dismissed from his position and expelled from the Communist Party in July 1986.
171-2.

Bronnikov, V.K.
Chief Engineer, Minsk nuclear power and heating plant.
159.
Bujak, Zbigniew.
Polish Solidarity underground activist.
66.
Buriak, V.N.
Deputy Chairman, Sanitary-Epidemiological Department, Ministry of
    Health of the Belorussian SSR.
162.
Burtica, C.
Minister of Trade, Romania.
83.
Ceasescu, Nikolai.
General Secretary of the Communist Party of Romania and President of
    Romania
58.
Champlin, Richard.
Physician at UCLA Medical Center. One of the Gale team at Moscow's
    Hospital No. 6.
138-9.
Chaus, D.D.
Head of the subunit of Kiev Security Guards' Department that guarded
    Prypiat apartments after the evacuation.
155.
Chazov, Evgenii.
Soviet Co-President of the International Physicians for the Prevention of
    Nuclear War.
176.
Chebrikov, V.M.
Chairman of the KGB.
174.
Chernenko, Konstantin.
Succeeded Iurii Andropov as General Secretary of the CC CPSU. Died in
    March 1985.
59.
Chernenko, L.
Correspondent, *TASS* news agency.
21.
Chernov, Boris.
Steam-turbine operator at Chernobyl nuclear power plant.
117.

Chnoupek, Bohuslav.
Foreign Minister, Czechoslovakia.
64.
Chumak, V.
Director of the Scientific Centre for Ecological Problems of Nuclear Energy, Academy of Sciences of the Ukrainian SSR.
102.
Churkin, Vitalii.
Second Secretary, Embassy of the USSR, Washington, D.C.
8.
Curin, Grisos.
Delegate to the Croatian Assembly, Yugoslavia. Opposed to the development of nuclear power in Yugoslavia.
68-9.
de Cuellar, Javier Perez.
Secretary General of the United Nations.
179.
Demichev, D.M.
First Party Secretary, Khoiniki Raion Committee, Gomel oblast, Belorussian SSR.
144-5.
Dinkov, V.A.
Minister of the Oil Industry of the USSR.
48-9.
Dmytruk, A.
Deputy Director, Construction Department, Khmelnytsky nuclear power plant. Severely reprimanded in 1983.
89.
Dolgikh, V.I.
Secretary, CC CPSU and Candidate Member of the CC CPSU Politburo.
104-5.
Dollezhal, N.
Soviet Academician who co-authored article opposing the development of nuclear power in the European part of the USSR.
100-1.
Domaniuk, Oleksander.
Party Secretary, Prypiat Urban Party Committee.
175.
Dubensky, Volodymyr.
Head of Construction Department, Odessa nuclear power and heating plant.
87.

Dubinin, Iurii.
Soviet Ambassador to the United States.
8, 179.
Duffy, Mike.
Correspondent, Canadian Broadcasting Corporation.
126.
Dzhur, A.I.
Brigadier, Odessa nuclear power and heating plant.
87.
Ehrenberger, Vlastimil.
Fuel and Power Minister, Czechoslovakia.
56.
Emelianov, Ivan.
Corresponding Member of the Academy of Sciences of the USSR and
Deputy Director of the Institute for Energy Technology of the
USSR. Involved with the design of the Chernobyl plant. Dis-
missed from his position as Deputy Director on 20 July 1986.
124-7, 164, 167, 170-1.
Falin, Valentin.
Director, *Novosti* press agency.
167.
Fazekas, Laszlo.
Hungarian correspondent.
124.
Feoktistov, Lev P.
Corresponding Member of the Academy of Sciences of the USSR and
Deputy Director of the Kurchatov Atomic Energy Institute of the
Academy.
111, 134.
Feshchenko, M.V.
Chief Engineer, Construction Department, Odessa nuclear power and
heating plant.
86.
Fiodorov, Viktor.
Minister of Petroleum Mining and Petrochemical Industry of the USSR
who was replaced in October 1985.
48.
Fomin, G.N.
Head of the USSR State Committee for Construction and Architecture
and First Deputy Chairman of the USSR State Construction
Committee. Lost both positions following the Atommash affair
in July 1983.
104.

Fomin, Nikolai.
Chief Engineer, Chernobyl nuclear power plant. Dismissed on 20 July
    1986 for alleged irresponsibility and inefficiency and sub-
    sequently expelled from the Communist Party.
117, 171-2.
Gale, Robert Peter.
Physician at the UCLA Medical Center. Chairman of the International
    Bone Marrow Registry in Milwaukee, Wisconsin. Volunteered
    help with bone marrow transplants for the most severely affected
    Chernobyl victims and headed international team at Moscow's
    Hospital No. 6. Has agreed to monitor progress of about 100,000
    Soviet citizens in future years.
33, 133, 137-40, 149.
Geddes, W.P.
Nuclear expert. Uranium Institute, London.
107.
Gerle, Ladislav.
Deputy Federal Premier, Czechoslovakia.
55.
Gorbachev, Mikhail.
General Secretary, CC CPSU since March 1985.
19, 24, 32-3, 46-8, 53, 73, 137, 167-8, 177-9.
Gratz, Leopold.
Foreign Minister, Austria.
64.
Gubarev, V.
Pravda correspondent.
13, 31.
Gusev, Vladimir.
Replaced Shcherbyna as head of the Government Commission investi-
    gating the accident.
154, 168.
Guskova, A.K.
Chief Radiologist, Hospital No. 6, Moscow. Led the Soviet and interna-
    tional teams performing bone marrow transplants on victims of
    severe radiation doses as a result of the Chernobyl accident.
136, 138-40, 149, 167.
Halkin, Dmytrii.
Minister of Ferrous Metallurgy of the Ukrainian SSR.
169.

Hammer, Armand.
Chairman of Occidental Petroleum. Arranged for Gale to travel to Moscow.
138.

Hanzhela, M.M.
Foreman, South Ukraine nuclear power plant.
81.

Havel, Stanislav.
Chairman of Atomic Energy Commission, Czechoslovakia.
63-4.

Hladush, I.D.
Minister of Internal Affairs of the Ukrainian SSR.
21, 32, 155.

Hora, V.T.
Chief Engineer, Chernobyl nuclear power plant.
121.

Hrotsenko, A.
Director of Construction Department, Khmelnytsky nuclear power plant.
89.

Hrynko, Mykola [Nikolai Grinko].
Minister of the Coal Industry of the Ukrainian SSR who was dismissed in October 1985.
39-40.

Hubskii, O.
Party member at Chernobyl nuclear power plant. One of the first officials to be dismissed after the accident.
169.

Ieltsin, Boris.
Candidate member, CPSU POlitburo and Chairman of Moscow City Party Organization.
8-11.

Ihnatenko, Evgenii.
Deputy Chairman of the Union of Atomic Energy industrial association, Ministry of Power and Electrification of the USSR.
159, 167.

Ihnatenko, Vasylii.
Fireman at Chernobyl nuclear power plant. Accident victim.
129-30, 140.

Ilin, Leonid.
Vice-President of the Academy of Medical Sciences of the USSR and Director of the Institute of Biophysics of the USSR.
22, 158, 176.

Itkin, V.
Correspondent, *TASS* news agency.
21.
Ivanovic, Dragisa.
Member of the Central Committee of the Communist Party of Yugos-
    lavia.
61.
Izrael, Iu.A.
Chairman of the USSR State Committee for Hydrometeorology and En-
    vironmental Control.
28-9, 133, 135, 173.
Jaruzelski, Wojciech.
General and Polish leader.
67.
Jarzebski, Stefan.
Minister of Environmental Protection, Poland.
12.
Kachura, Borys.
Secretary of the Politburo of the Communist Party of Ukraine.
76, 174.
Kapitsa, Petr.
Soviet Academician.
98.
Karpenok, M.I.
Elderly Prypiat resident who hid during the evacuation of the town.
145.
Kasianenko, A.M.
Deputy Minister of Health of the Ukrainian SSR.
149.
Keleberda, U.
Army Colonel. Had major responsibilities during the clean-up operation
    after the accident.
155.
Khodemchuk, Valerii.
Operator at the Chernobyl plant's fourth unit. One of two people be-
    lieved to have been killed instantly after the accident, but his
    body was never recovered.
32, 137.
Kibenok, Viktor.
Twenty-three-year old fireman who was one of the first to fight the fire
    that broke out at the fourth unit of the Chernobyl plant. Died of
    an overdose of radiation fifteen days later.
13, 129-30, 140.

Kizima, V.T.
Head of Construction Department, Chernobyl nuclear power plant. Had almost sole responsibility for the plant in its earlier years.
99. 118-21.
Klebanov, Vladimir.
Former shift foreman in the Donetsk coalfield.
39.
Kolesnyk, S.M.
Electrician, South Ukraine nuclear power plant.
82.
Kolinko, Vladimir.
Correspondent, *Novosti* news agency. Provided one of the first detailed accounts of the accident scene.
16-18.
Kordyk, Z.F.
Head of the Chernobyl Meteorological Station.
13.
Koriakin, Iu.
Doctor of economic sciences. Co-authored article opposing nuclear energy development in European part of the USSR.
100-1.
Kovalev, Anatolii.
First Deputy Minister of Foreign Affairs of the USSR.
142.
Kovalevska, Liubov.
Editor of the local Prypiat newspaper. Author of major critique of the Chernobyl plant in *Literaturna Ukraina*.
122-3.
Koziakov, Vladislav.
Party spokesman on *Radio Moscow*
178.
Kozlov, Nikolai.
Deputy Chairman, USSR State Committee for Hydrometeorology and Environmental Control.
133.
Krasnikov, M.I.
Building Engineer, Chernobyl nuclear power plant.
116.
Krutov, Mikhail.
Reporter, Moscow television.
11, 16, 29-30.

217

Kulic, Slavko.
Scientific Advisor at the Zagreb Economic Institute and head of the Centre for Strategic Research, Yugoslavia.
61.
Kulov, Evgenii.
Chairman of the USSR State Committee for the Supervision of the Safe Conduct of Work in the Nuclear Power Industry. Appointed at the committee's inception in August 1983. Dismissed as a result of the Chernobyl accident on 20 July 1986.
105, 171.
Kurgan, M.
Head of the section of South Energy Construction Isolation at Khmelnytsky nuclear power plant. Put on trial in 1983 for alleged theft of funds.
89.
Kurguze, A.
Operator at Chernobyl nuclear power plant. Accident victim.
140.
Kushnir, A.
Head of a section of South Energy Construction Mechanization at Khmelnytsky nuclear power plant. Put on trial in 1983 for alleged embezzlement of construction materials.
89.
Lapko, A.
Secretary of the party committee supervising construction work at Khmelnytsky nuclear power plant. Severely reprimanded in 1983.
89.
Legasov, Valerii.
Member of the Presidium of the Academy of Sciences of the USR and Deputy Director of the Kurchatov Atomic Energy Institute of the Academy.
29, 112-13, 150, 176.
Legun, S.
Fireman at the Chernobyl nuclear power plant.
130.
Lelechenko, ?.
Plant worker at Chernobyl nuclear power plant. Accident victim.
140.
Liashko, Oleksander.
Chairman of the Council of Ministers of the Ukrainian SSR.
9, 23-4, 39, 142, 167-8.

Ligachev, Egor.
Secretary of the CC CPSU Politburo.
9-10, 20, 24, 144, 155, 167-8.
Lisovsky, T.
First Party Secretary, Khmelnytsky Oblast Party Committee.
89.
Lomeiko, Vladimir.
Spokesman for Foreign Ministry of the USSR.
8.
Lopachev, A.
Deputy Chief of Construction Department at Khmelnytsky nuclear power
    plant. Reportedly embezzled materials.
89.
Lupyi, Halyna.
Komsomol leader, Food Supply Department, Chernobyl nuclear power
    plant. Allegedly fled from the station after the accident.
169.
Maiorets, A.I.
Succeeded P. Neporozhny as Minister of Power and Electrification of the
    USSR in the summer of 1985. Severely reprimanded after the
    Chernobyl accident.
53, 74, 171.
Makukhin, O.N.
First Deputy Minister of Power and Electrification of the USSR and for-
    mer Minister of Power and Electrification of the Ukrainian SSR.
    Severely reprimanded in August 1986 for failure to improve reli-
    ability of Chernobyl plant.
81.
Marchuk, H.
Deputy Chairman of the Council of Ministers of the USSR and Chairman
    of the USSR State Committee for Science and Technology.
98.
Marsham, Tom.
British nuclear expert.
109.
McCally, Michael.
Professor of Clinical Medicine at the University of Chicago. Member of
    the Gale team in Moscow.
139.
McReynolds, James S.
Leader of a group from the U.S. Episcopal Church visiting Odessa in
    May 1986.
22.

Medvedev, Zhores.
Exiled Soviet geneticist who helped to publicize the Urals nuclear disaster in the West.
103, 113.
Mernenko, V.M.
Second Secretary, Chernobyl Raion Party Committee.
144.
Michel, Robert H.
U.S. Representative from Illinois who met with P. Neporozhny in 1979.
113.
Mikheiev, V.
Correspondent, *Izvestiia*.
96.
Nekrasov, A.M.
Soviet economist.
95.
Neporozhny, Petro.
Former Minister of Power and Electrification of the USSR. Dismissed in 1985.
73, 113, 121.
Nikiforov, G.A.
First Party Secretary, Ekibastuz City Party Committee.
42.
Novikov, I.
Chairman of the USSR State Committee for Construction who went into retirement after the Atommash affair in 1983.
104-5.
Nychyporenko, M.
Fireman, Chernobyl nuclear power plant.
130.
Odinets, M.
Correspondent, *Pravda*.
13, 31.
Odintsov, V.
Head of an Assembly Department, Khmelnytsky nuclear power plant. Put on trial for stealing rolled sheet metal in 1983.
89.
Palous, Milos.
Member of Charter 77 movement, Czechoslovakia.
63.
Patolichev, N.
Minister of Trade of the USSR.
83.

Pavlov, Igor.
Commentator, *Radio Moscow.*
5.
Petrosiants, A.M.
Chairman of the USSR State Committee for the Utilization of Nuclear
    Energy.
97, 109, 177, 179.
Petrov, Alexander.
Deputy Chairman of the Council of Ministers of the Belorussian SSR.
    Head of the working group of the clean-up operation in Belorus-
    sia.
161.
Petrovsky, A.
Fireman, Chernobyl nuclear power plant.
129-30, 139-40.
Petrushenko, V.
Brigadier at Khmelnytsky nuclear power plant. Put on trial for theft of
    rolled sheet metal in 1983.
89.
Pimenenko, Ia. N.
Head of the Kiev Centre for the Study and Monitoring of the Environ-
    ment.
28.
Pisarevsky, V.
Correspondent on Soviet television.
159.
Plaksienko, A.V.
Brigadier, Odessa nuclear power and heating plant.
86.
Planinc, Milka.
Premier, Yugoslavia.
61.
Pliushch, I.S.
Chairman of Kiev Oblast Executive Committee.
168.
Plokhy, T.G.
Former Deputy Chief Engineer at Chernobyl nuclear power plant, sub-
    sequently moved to Balakovo nuclear plant.
159.
Pozdyshev, E.
Appointed Director of Chernobyl nuclear power plant, July 1986.
171.

Pozmogov, Anatolii.
Physician and Director of Kiev Radiology Institute.
140.
Pravyk [Pravik], V.P.
Fireman at Chernobyl nuclear power plant. Accident victim.
13, 129-30, 140.
Pryshchepa, V.
Fireman at Chernobyl nuclear power plant.
130.
Reisner, Yair.
Israeli biophysicist. Member of Gale's team in Moscow.
138.
Revenko, Hryhorii.
First Party Secretary of Kiev Oblast Committee.
9, 20, 27, 147, 155, 168-9.
Romanenko, A. Iu.
Minister of Health of the Ukrainian SSR.
22, 28, 148.
Romanets, Anatolii.
Official, Ministry of Health of Ukrainian SSR.
19.
Rosen, Morris.
Head of the Division of Nuclear Safety, IAEA.
26-7.
Rylsky, Maxim.
Soviet journalist.
117.
Ryzhkov, Nikolai.
Chairman of the Council of Ministers of the USSR.
9-10, 24, 144, 155, 167-8, 174, 177.
Ryzhov, Mikhail.
Member of Soviet delegation to IAEA.
179.
Saakov, E.S.
Chief Engineer with an association of the Ministry of Power and Electrification of the USSR.
159.
Sabatova, Anna.
Member of Charter 77 movement, Czechoslovakia.
63.
Scheer, Jens.
West German nuclear physicist.
117.

Schultz, George.
Secretary of State, United States government.
137.
Seledovkin, G.D.
Physician, Moscow Hospital No. 6.
137.
Semeniaka, A.S.
Elderly Prypiat resident who hid during the evacuation of the town.
145.
Semeniuk, V.M.
Former Minister of Power and Electrification of the Ukrainian SSR.
38.
Semenov, Boris.
Soviet representative to the Board of Governors, IAEA. Deputy Chairman of the State Committee for the Utilization of Nuclear Energy.
125, 136, 178-9.
Shapoval, A.
Leader of Chernobyl subsidiary of the Southern Atomic Energy Construction Association. One of first to be dismissed after Chernobyl accident.
31, 169.
Shasharin, G.
First Deputy Minister of Power and Electrification of the USSR. Dismissed as a result of Chernobyl accident on 20 July 1986.
98-9, 103, 171.
Shashenok, Vladimir.
An adjustor of automatic systems at the Chernobyl nuclear power plant. Died instantly after the accident from burns.
32, 137.
Shavrei, Ivan.
Fireman, Chernobyl nuclear power plant. Played major role in extinguishing blaze on reactor roof.
129-30.
Shchadov, M.I.
Minister of the Coal Industry of the USSR since December 1985. Supervised digging of tunnel under damaged Chernobyl reactor.
164-5.
Shchekin, Anatolii.
Chairman of Chernobyl Raion Executive Committee.
172.

Shchepin, Oleg.
First Deputy Minister of Health of the USSR.
136.
Shcherbany, A.
Deputy Director, Construction Department, Khmelnytsky nuclear power plant. Severely reprimanded in 1983.
89.
Shcherbyna, Borys.
First head of the Government Commission set up to investigate the consequences of the Chernobyl accident. A Ukrainian, born in 1920, he was First Party Secretary in Tiumen Oblast in western Siberia during the boom years of the oil industry. Believed to have fallen ill during his work at the accident site.
4, 9, 14-17, 19, 21, 23-4, 124, 154, 157, 168.
Shcherbytsky, Volodymyr.
Member of the CC CPSU Politburo and First Party Secretary of the Communist Party of Ukraine.
7, 24, 85, 147, 154-6, 168.
Shenydlin, A.
Director, High Temperature Institute, Academy of Sciences of the USSR.
100.
Shevchuk, S.M.
Head of the Rovno nuclear power plant section of the Chernobyl Energy Association.
80.
Shpak, I.F.
Chief Engineer, Crimea nuclear power plant.
90.
Sichkarenko, A.
Leader of collective of the Chernobyl subsidiary of South Energy Construction Transport Association. Dismissed on 12 May 1986 after Chernobyl accident.
31, 169.
Silaiev, Ivan.
Deputy Chairman, Council of Ministers of the USSR.
29-30, 154, 158-9, 164.
Skliarov, Vitalii.
Minister of Power and Electrification of the Ukrainian SSR.
76, 78, 80.
Sliunkov, N.N.
First Secretary, Communist Party of Belorussia.
156, 161.

Sokolov, Vladimir.
Kiev correspondent of *Radio Moscow*.
15, 24, 150.
Sokolovsky, Valentin.
Deputy Chairman of the USSR State Committee for Hydrometeorology and Environmental Control.
12.
Soloviev, R.
Deputy Director of Chernobyl nuclear power plant. Reportedly abandoned his post after the accident.
171.
Sosedenko, O.
Head of Construction Department, South Ukraine nuclear power plant.
81.
Sowinski, Mieczyslaw.
Chairman, Polish Nuclear Agency.
67.
Speka, P.M.
Foreman, South Ukraine nuclear power plant.
81.
Stalin, I.V.
Soviet leader and General Secretary CC CPSU from the late 1920s until death in 1953.
20, 105.
Staroshchuk, Slava.
Party member at Chernobyl nuclear power plant. Reportedly fled to Odessa after the accident.
170.
Stepanenko, N.
Deputy Chairman, Kiev Oblast Executive Committee.
170.
Stern, Jan.
Member of Charter 77 movement, Czechoslovakia.
63.
Surgai, N.
Minister of the Coal Industry of the Ukrainian SSR since October 1985.
164.
Sychev, V.
Secretary of the CMEA.
53.

Sydorenko, Viktor.
Deputy Chairman of the USSR State Committee for the Supervision of Safe Practices in the Nuclear Power Industry. Dismissed from his post in August 1986.
125, 167, 176.
Szalada, Zbigniew.
Deputy Premier of Poland. Head of Polish Commission set up after the Chernobyl accident.
12.
Targan, Robert.
Hungarian energy planning official.
68.
Teliatnikov, Leonid.
Fire Chief at Chernobyl nuclear power plant. Graduate of Sverdlov firemen's technical school, born in 1951. Hospitalized in Moscow after putting out the fire. Seriously injured, but subsequently released.
25-6, 129-30, 139-40.
Terasaki, Paul.
Specialist in tissue typing at the UCLA Medical Center. Member of the Gale team in Moscow.
138.
Tishchura, Volodymyr.
Fireman at Chernobyl nuclear power plant. Accident victim.
129-30, 140.
Titenok, N.
Fireman at Chernobyl nuclear power plant. Accident victim.
130, 140.
Tito, Josip.
Late Yugoslav leader and statesman.
69.
Todoriev, Nikolai.
Minister of Power Engineering, Bulgaria. Corresponding Member of the Bulgarian Academy of Sciences.
54-5, 67.
Tokarasky, Boris.
Soviet emigre living in Israel. Involved in the construction of Chernobyl nuclear power plant before 1978.
120.
Toporkova, T.T.
Physician, Hospital No. 6, Moscow.
137.

Troitsky, A.A.
Soviet economist.
95.
Tsarenko, I.
A Deputy Director at Chernobyl nuclear power plant. Dismissed on 20 July 1986.
171.
Tytarenko, O.O.
Second Secretary, Communist Party of Ukraine.
7.
Urban, Jerzy.
Polish government spokesman.
65-6.
Vashchuk, Mykolai.
Fireman, Chernobyl nuclear power plant. 129-30, 140.
Vasiliev, A.
Police colonel on patrol in Chernobyl.
145.
Vazhenov, E.
Deputy Chief of Construction Department, Khmelnytsky nuclear power plant.
89.
Vedernikov, G.G.
Deputy Chairman, USSR Council of Ministers. In August, was appointed the head of the Government Commission investigating the accident.
154.
Velikhov, Evgenii P.
Vice-President, Academy of Sciences of the USSR. Scientific advisor to Mikhail Gorbachev. Supervises the Commission on Energetics of the USSR Supreme Soviet. Became a Candidate Member of the CPSU Central Committee in March 1986. The chief scientist involved in the Chernobyl clean-up campaign.
21, 29, 126, 142, 157, 160, 172-3.
Vereshchaka, P.I.
Foreman, South Ukraine nuclear power plant.
81.
Veretennikov, G.
Deputy Minister of Power and Electrification of the USSR. In August 1986, was expelled from the Communist Party as a result of the Chernobyl accident.
89, 97.

Vetchinin, V.V.
Chairman of the Epidemiological Department, Ministry of Health of the Ukrainian SSR.
158.
Voronin, Lev.
Deputy Chairman of the USSR Council of Ministers and Chairman of the USSR State Committee for Material and Technical Supply. Replaced Ivan Silaiev as the chief official of the clean-up campaign at the accident site.
154.
Walker, Christopher.
Correspondent, *The Times*, London.
148.
Zahalsky, Iurii.
Deputy Head of the youth section of the Construction Department at Chernobyl nuclear power plant. Allegedly "shirked his duties" after the accident.
169.
Zahorulko, I.I.
Second Secretary, Rovno Oblast Party Committee.
80.
Zelinsky, A.N.
First Deputy Minister of Health of the Ukrainian SSR.
23.
Zhivkov, Todor. '
First Party Secretary of Bulgaria.
53.
Zhukovsky, V.
Correspondent, *TASS* news agency.
21.
Zolton, Iurii.
Commentator on *Radio Moscow*.
5.